AF332993

Automotive Refinement

Autotech 95 Organizing Committee

C Bale
Tickford Limited

I Birchall
Prodrive Engineering Limited

J Chelsom
City University

M Dunn (Chair)
Consultant

P Extance
Lucas Advanced Engineering Centre

C Hickman
MIRA

R Johnson
SDRC Engineering Services

P Marshall
Consultant

I Noble
Consultant

S Ocock
Rover Group Limited

B Southall
Perkins Technology Limited

D Tidmarsh
University of Central England

J Turner
University of Southampton

M Williams
Jaguar Cars Limited

IMechE
Seminar Publication

Automotive Refinement

Selected papers from Autotech 95
7–9 November 1995

Organized jointly by the Automobile Division
of the Institution of Mechanical Engineers (IMechE)

and in association with the Institution of Electrical Engineers

IMechE Seminar Publication 1996 – 7

Published by Mechanical Engineering Publications Limited for
The Institution of Mechanical Engineers, London.

First Published 1996

© The Institution of Mechanical Engineers 1996

ISSN 1357-9193
ISBN 1 86058 021 1

A CIP catalogue record for this book is available from the British Library
Printed by The Ipswich Book Company, Suffolk, UK

Contents

Related Titles of Interest

Title	Author	ISBN
Automotive Manufacturing	IMechE Seminar 1996–4	1 86058 023 8
Automotive Systems, Sensors, and Signalling	IMechE Seminar 1996–5	1 86058 022 X
Automotive Passenger Safety	IMechE Seminar 1996–6	1 86058 042 4
Automotive Powertrains	IMechE Seminar 1996–8	1 86058 020 3
Set of 5 Autotech Volumes (including the 4 titles listed above and this title)	IMechE Seminar 1996	1 86058 019 X
Handbook of Vehicle Design Analysis	J Fenton	0 85298 963 6
Application Of Power Train And Fuel Technologies To Meet Emission Standards	IMechE Conference 1996–5	0 85298 996 2
Diesel Fuel Injection Systems	IMechE Seminar 1995–3	0 85298 980 6
Using Natural Gas in Engines	IMechE Seminar 1996–2	1 86058 033 5

For the full range of titles published by MEP contact:

Sales Department
Mechanical Engineering Publications Limited
Northgate Avenue
Bury St Edmunds
Suffolk
IP32 6BW
UK

Tel: 01284 763277
Fax: 01284 704006

Noise and Vibration

Modelling absorbent materials in acoustic prediction and diagnostics

C F McCULLOCH MA, CEng, MIMechE, MIOA
Dynamic Structures and Systems Limited, UK

SYNOPSIS

The vibro-acoustic behaviour of multi-layered composite materials used to reduce noise and vibration is modelled using a finite element approach. An advanced model is used for the poro-elastic materials, taking into account the acoustic and structural interactions in the interior of such materials. The material parameters of this model can be derived from measurements. The detailed model of the behaviour of the complete multi-layered system can be linked to analyses of the vibro-acoustic behaviour of a larger assembly or the whole vehicle. Some practical applications of the approach related to vehicle refinement are discussed.

NOTATION

p	pressure in fluid
u	displacement of fluid
U	displacement of poro-elastic frame
ρ_s	density of fluid
ρ_f	density of poro-elastic frame
Ω	porosity
R	resistivity
α_{oo}	tortuosity
α	scaling factor (Biot's α)
Q	bulk modulus of fluid in pores
E	Young's modulus
ν	Poisson' ratio
Λ	characteristic viscous length
Λ'	characteristic thermal length
$L^T\sigma$	stress gradient (tensor)
ε_s	volumetric strain of poro-elastic skeleton
ε_v	net fluid flow into unit volume of poro-elastic/fluid mixture

1 INTRODUCTION

In recent years, methods for predictive acoustic calculations have been developed, particularly those based on finite element (FE) and/or boundary element (BE) models of the behaviour of the acoustic fluid. These methods have gained widespread acceptance for many applications, in the automobile and other vehicle industries, aerospace, and other branches of engineering where vibro-acoustics is significant. Typically, these FE and BE models of the acoustic behaviour (*ie* acoustic pressure waves in the fluid) have been linked with FE models of structural behaviour, either taking the structural vibration data as boundary conditions of the acoustic behaviour, or performing fully-coupled analyses in which an FE description of the structure interacts with, and has its behaviour modified by, an FE or BE model of the associated fluid. Discussion of these approaches, which are now regarded as standard, can be found in the literature [1,2,3].

However, the acoustic damping provided in all such models is usually limited: it may be defined only as surface impedances (complex, possibly frequency-dependent) or possibly by volume absorption behaviour (typically, a porous acoustic material, with rigid fibres or cells, which are considered to be stationary: the material is characterized by a flow resistance, a porosity, and a structure-factor defining how tortuous is a typical path from cell to cell).

Such descriptions may be inadequate for more complicated assemblies and composite or laminated materials: the 'global' material properties may in any case not be known, or it may be desirable to tune the behaviour of a built-up material to optimise performance.

This paper describes an approach which has been developed to enable such complicated built-up materials to be modelled and analyzed, taking into account the multiple internal fluid-structure interactions which can take place. The methods are embodied in a computer program which has been made as general as possible to facilitate its use in many different applications. A few of these are described, with some practical data, in the latter part of the paper.

2 CHARACTERISTICS OF PORO-ELASTIC MATERIALS

Sandwich panels consisting of three, four or more layers are widely used in the automotive and other industries, to control noise and vibration. Typically, they can have layers of four types:

- a sub-strate (structural) layer, usually sheet metal
- a damping layer, applied to the sub-strate
- a porous layer, usually relatively thick
- a heavy visco-elastic (or 'limp') layer.

Air gaps may also be present, maintained by relatively widely-separated spacers, and the interfaces between the layers may have adhesive with through-thickness and tangential (shear) flexibilities. Such an array is shown schematically in Figure 1. Real assemblies may, of course, not have examples of all the above types or may have many more layers than four in all.

The multi-layer construction acts firstly, in a 1-dimensional sense, through the thickness, as a system of masses and springs. The sheet metal sub-strate and the heavy layer act as masses and the springs are defined by the stiffness of the skeleton of the poro-elastic material, its resistivity (pressure drop for a given fluid flow velocity) and compressibility of any fluid trapped inside closed pores. Below a cut-off frequency, vibrations due to mechanical excitation may be amplified between the sub-strate and the outer surface, and above this frequency they will be reduced.

Secondly, viscous damping in all the layers dissipates energy. The level of damping is usually small in the sub-strate (*eg* steel) and in the porous layers (since they contain little structural material) but is usually significant in the damping layer and in the limp layer.

Thirdly, acoustic flow (cyclic particle motion) in the porous layers causes a dissipation of energy due to the relative motion between the fluid and the skeleton of the material. This occurs in the 1-dimensional sense, through the thickness but can be especially significant when 2- or 3-dimensional behaviour is considered: lateral flows can then be set up, due to out-of-phase motions of layers above and below the porous one, greatly increasing the dissipation of energy. This can also be considered in terms of mis-matched bending wavelengths in the upper and lower layers (see Figure 2).

3   MATHEMATICAL MODELS

3.1 Poro-elastic model

The porous material can be viewed as a two-phase mixture, with fluid within the pores and the skeleton of the material forming an elastic frame. Note that in this case, only open pores (which communicate with one another) are of interest: closed pores are taken into account only in the sense that they modify the behaviour of the elastic frame. The interactions between the two material phases are described by the theory due to Biot [4,5] which will be summarized here. The main assumptions are:

-    acoustic wavelengths are much greater than the details of the poro-elastic materials
-    displacements and pressure perturbations are small (*ie* linear elastic behaviour)
-    the fluid phase is continuous (the pores communicate with each other, see above)
-    the skeleton is elastic (viscous effects due to fluid in closed pores are neglected)
-    thermo-mechanical coupling effects are neglected (the fluid behaves adiabatically).

The Biot theory is based on a model of the observed physical phenomena in the behaviour of such materials, rather than any micro-physical description.

An overall equilibrium equation can be set up for the mixture, and an equation based on the generalized Darcy law for the fluid phase. Constitutive relations can be written for the mixture (related to total stresses) and for the fluid phase (related to the pressure). Appropriate boundary conditions are then added to complete the equation system.

Defining the state of the system by three initial variables, p the pressure in the fluid, u the displacement of the frame and U the displacement of the fluid, we derive the relative displacement of the fluid with respect to the frame as U - u. Note the difference between this displacement formulation and the conventional finite element model for purely acoustic problems, in which it is more convenient to use the fluid particle velocity: using displacement leads to a homogeneous formulation for the coupled problem.

The material is defined by ρ_s the density of the frame, ρ_f the density of the fluid, the porosity Ω, the resistivity R, the tortuosity α_{∞} and the bulk modulus of the fluid in the pores Q, together with the conventional Young's modulus E and and Poisson's ratio ν. (R and Q are in general frequency-dependent and can be related to the characteristic viscous and thermal lengths Λ and Λ', using for example the model of Allard and Johnson [6]. Space precludes a detailed discussion of this). Methods to determine these materials properties will be discussed later in the paper. If we further represent the stress gradient in the material by $L^T\sigma$, we can derive an equation for the equilibrium of the mixture in which external forces balance inertia forces:

$$L^T\sigma - (1 - \Omega)\,\rho_s\,u - \Omega\,\rho_f\,U = 0 \tag{1}$$

where $(1 - \Omega)\,\rho_s\,u$ is the acceleration of the frame and $\Omega\,\rho_f\,U$ is the acceleration of the fluid. The equilibrium of the fluid alone is defined in terms of the pressure gradient ∇p, viscous forces due to the resistivity R (U - u), inertial forces $\rho_f\,U$, and a mass coupling term $\rho_f\,(\alpha_{\infty} -1)(U$ - u) due to the tortuosity (the path from one cell to the next is not direct). Hence:

$$\nabla p + R\,(U - u) + \rho_f\,U + \frac{\rho_a\,(U - u)}{\Omega} = 0 \tag{2}$$

writing $\rho_a = \rho_f\,\Omega\,(\alpha_{\infty} -1)$ as the effective density of the fluid.

A constitutive law can be written for the mixture, as:

$$\sigma = D\,\varepsilon - \alpha\,m\,p \tag{3}$$

where σ is the stress tensor, D is a conventional Hooke matrix, and m is a vector to convert the scalar pressure p into an isotropic vector. α (not to be confused with tortuosity, α_{∞}) is sometimes referred to as Biot's α, and is a scaling factor, normally 1, which can be adjusted to obtain a better correlation with empirical data in some cases. Note that the sign convention for stresses (positive tensile) and pressures (positive compressive) results in a negative sign on the right-hand side of (3).

A constitutive law can be written for the fluid, as:

$$-p = \alpha\,Q\,\varepsilon_s + Q\,\varepsilon_v \tag{4}$$

where ε_s is the volumetric strain of the skeleton and ε_v is the net fluid flow (positive inwards) per unit volume of mixture. The latter term can thus take into account leakage of fluid from the boundaries of the poro-elastic region, which may be defined to be open or closed (sealed) as appropriate.

Equations (1) and (2) can be modified to take into account the effects of porosity Ω, by writing relative displacement $w = \Omega(U - u)$. This enables layers with different porosities to be handled straightforwardly, without any special treatment at the interfaces: compatibility (continuity of fluid volume velocity) is automatically satisfied.

Applying a standard virtual work principle and finite element approach leads to a finite element model for the porous material based on the above equations, with the usual benefits from an FE model of symmetric and banded matrices. Element functions based on volume elements such as 8-noded hexahedra are used.

3.2 Structural layers: hybrid shell element

The structural layers (non-porous) of the multi-layered assembly are modelled using conventional finite elements, with standard properties. (Young's modulus and Poisson's ratio are complex where damping losses are to be included in these layers).

However, for shell structures, it is conventional to use a thick or thin shell formulation for a standard finite element dynamic analysis, but this leads to problems in coupling to the volume elements used for the thicker layers such as the poro-elastic material. (There are six degrees of freedom at the nodes of the shell elements, including rotations, but only three at the nodes of the volume elements). This is overcome by using a hybrid shell element, which is topologically the same as a volume element, but has a formulation able to cope with the severe aspect ratios (length/thickness) which are normally unacceptable for standard volume elements.

3.3 Solution procedures

The FE equation system can finally be solved in the frequency domain, usually stepping through a range of frequencies, with a direct solution of the equations at each frequency, or in the time domain (transient) using for example Newmark's integration approach.

4 MEASUREMENTS OF MATERIAL PROPERTIES

In the preceding presentation, various parameters were introduced to describe the behaviour of the materials in the multi-layered construction. The conventional properties of structural parts are usually reasonably well known, or can be established by mechanical tests. However, the properties of the poro-elastic materials are less well-known. Various measurements can be performed:

Porosity (Ω) can be found by slowly compressing the porous material until all fluid is expelled, and comparing the initial and final volumes. An alternative is to saturate the pores with liquid and measure the volume of liquid which can be contained within a given volume of the bulk material.

Resistivity (R) can be measured as the pressure drop through a given thickness of material with a given flow rate of the fluid: $R = (p1 - p2) / h.v$

Tortuosity (α_{oo}) can be found at high frequencies by simply inserting a sample of the material between a miniature loudspeaker (ultrasonic source) and a microphone: the sound speed is determined solely by the tortuosity, so this can be found immediately from an impulse or similar test. More generally, tortuosity is related to the mean path length from cell to cell, hence it can be found by saturating the porous material with a conducting fluid and comparing the electrical resistance of the mixture with the resistance of the fluid alone along a direct path.

Biot's scaling parameter (α) for adjusting coupling terms, can be found by comparing measurements in a standard impedance tube with predictions from numerical models of the material in such a tube, without any adjustment of α (*ie* $= 1$) and with all other parameters adequately known. Any necessary adjustment of α can then be determined.

The structural properties of the skeleton of the poro-elastic material may also not be well known, compared to conventional materials. Some other, non-porous, layers may also have exotic materials. In all cases, the dynamic properties must be established. This can be done on a dynamic test rig in which a sample of the material is excited axially or torsionally. Accelerometers attached at the top and the bottom give transfer functions of the dynamic response. From the axial data, the dynamic Young's modulus (E) is found, and from the torsional data, the dynamic shear modulus (G) is found, from which Poisson's ratio (v) can be derived. (Figure 3).

5   SOFTWARE IMPLEMENTATION

The modelling of a general multi-layered vibro-acoustic system using the principles outlined above has been implemented into a computer program [7]. The program allows any number of layers in the construction. The geometry is arbitrary: the user may select any number of elements per layer (depending on the detail of the results and the mesh refinement required) which can be 1-dimensional (a 'stack' of elements) or 2- or 3-dimensional. Fully 3-dimensional geometry can have arbitrary interface surfaces between layers (complex, non-planar...). Interfaces to standard mesh generators allow the geometry to be loaded from a file in industry-standard formats.

The interfaces between porous and other layers are automatically impervious and the user may additionally define connections between layers with normal and tangential flexibilities (*ie* thin layers of 'glue'). The side faces of the model can be either open or sealed (impervious) to represent cases where fluid is allowed to 'breath' in and out. Boundary conditions can be defined as mechanical loadings (forces on the structural parts) or acoustic loadings (incident pressures).

All definitions of data and manipulations of the model are carried out within a Motif-based graphical user interface, which also provides visual checks of boundary conditions, sets of elements (material definition in layers) and presentation of results. The solvers use optimised algorithms, either in-core or (for larger problems) out-of-core, in the latter case with an automatic selection of the optimum block size.

Results can be derived as frequency (or time) response functions of acoustic pressures, velocities and intensities, and impedance (or admittance) and structural displacements. Energy checks are made at each frequency or time step, and the strain energy, kinetic energy, loss energy, work of interaction and resistivity forces in or between each layer can be found. Where a frequency-dependent surface impedance has been derived, it can be transferred and used in other software [8], or where surface vibrations have been derived, from known forcing on the sub-strate, these vibrations can be mapped onto an acoustic radiation model using FE or BE to define the boundary conditions of an acoustic calculation.

6 APPLICATIONS

Some applications of the software will be described, to show typical examples of its use. The main areas of application are:

- to compare the performance of different composite materials
- to investigate alternative fabrication strategies
- to optimise treatments intended to improve noise and vibration behaviour
- to calculate the excitation from treated surfaces, for noise radiation prediction.

6.1 Derivation of surface impedance

This example is based on a material used for the roof lining of a certain car. The material has two layers of foam (polycarbodiimide and polyurethane) covered by a facing material and bonded to a rigid backing (the structure). A layer with a high resistivity separates the foam layers from the facing. The aim of the calculation is to derive the surface impedance seen by the main passenger cavity of the car, which can then be used in a subsequent analysis of the interior acoustic behaviour.

The model is shown in Figure 4, with a coarse and a fine mesh. Note that the calculation is 1-dimensional, using a 'stack' of elements through the thickness. The four sides of this stack are made impervious. Structural and acoustic material properties for the layers are given in Table1. Figure 5 shows the predicted impedance compared to some measured values, with which a very good agreement can be seen.

6.2 Derivation of transmission loss of a composite panel

It is often of interest to know the transmission loss of a multi-layered panel, such as the firewall bulkhead in a car, between the engine bay and the passenger compartment. In the simplified example shown here, a poro-elastic layer is bonded onto an elastic layer. The incident sound field is considered to arrive as plane waves at normal incidence, hence again a 1-dimensional model can be used. Materials data are given in Table 2. The transmission loss is taken to be the ratio of the pressures between the two outside faces, and is shown graphically in Figure 6.

6.3 Derivation of surface vibrations and calculation of noise radiation

In this example, a simplified 3-dimensional model is used to show the derivation of noise radiation from surface vibrations, which are caused by forces on the sub-strate layer transmitted through a poro-elastic layer and a heavy surface layer. Symmetry allows only one quarter of a plate to be modelled, with appropriate boundary conditions (structural and acoustic) on the symmetry planes and at the outside edges of the model. Note also that the interfaces between the layers are made impervious. The materials data are given in Table 3.

The results for the multi-layered model at different frequencies are shown in Figure 7. It can be seen that at lower frequencies the sub-strate and the heavy surface layer tend to move together and there is not so much dissipation of energy in the porous layer. At higher frequencies, the upper and lower layers move with differing amplitudes and may be out of phase, and there is a much greater dissipation of energy.

7   CONCLUSIONS

A method for modelling the vibro-acoustic behaviour of multi-layered constructions including poro-elastic materials has been described.

The method is embodied in a computer program which allows simple and complex calculations to be performed on standard computers in a robust environment for the user.

Many applications of the method can be foreseen, for deriving and optimising the surface absorption behaviour of acoustic treatments, assessing the transmission loss of layered panels, and predicting the noise radiation from treated structures.

ACKNOWLEDGEMENTS

The assistance of various colleagues in preparing material for this paper, especially Dr J-P Coyette and Mr J-L Migeot of Numerical Integration Technologies and Dr H Bloemhof of Rieter, is gratefully acknowledged.

REFERENCES

[1] von Estorff, O 'Numerical Treatment of Acoustic Problems - an Overview' 10th Intl FE Congress, Baden-Baden, Germany, 1991

[2] McCulloch, C F 'Practical methods for predictive acoustic modelling' IMechE Autotech'93, C462/3/044, Birmingham, UK, 1993

[3] Theory Manual SYSNOISE, Numerical Integration Technologies, Leuven, Belgium, 1995

[4] Biot, M A 'Theory of propagation of elastic waves in a fluid-saturated porous solid' Jnl Ac Soc Am, Vol 28 pp 168-191, 1956

[5] Biot, M A 'Mechanics of deformation and acoustic propagation in porous media' Jnl Appl Phys, Vol 33 pp1482-1498, 1962

[6] Allard, J-F 'Propagation of sound in porous media (modelling sound-absorbing materials)' Elsevier, 1993

[7] Users Manual VIOLINS, Numerical Integration Technologies, Leuven, Belgium, 1995

[8] *eg* SYSNOISE, *eg* see [3]

TABLE 1:   **Data for surface impedance prediction**

Mechanical properties:

Layer	thickness (m)	ρ_s (kg/m3)	E (kN/m2)	ν
1	0.004	2050.	110 + 1.6j	0.3
2	0.0008	625.	1000 + 100j	0.3
3	0.005	1033.	55 + 3.0j	0.3
4	0.016	1600.	18 + 1.8j	0.3

Porous (acoustic) properties:

Layer	Ω	α_{∞}	R (kNm⁻⁴s)	Λ (10⁻⁶m)	Λ' (10⁻⁶m)
1	0.98	1.18	34.	60.	86.
2	0.80	2.56	3200.	6.	24.
3	0.97	2.52	87.	36.	119.
4	0.99	1.98	37.	37.	121.

TABLE 2:   **Data for layered material transmission loss prediction**

Mechanical properties:

Layer	thickness (m)	ρ_s (kg/m3)	λ (GN/m2)	μ (GN/m2)
1	0.0016	2800.	56 + 0.7j	24 + 0.4j
2	0.0257	1666.	0	0.01 + 0.0001j

Porous (acoustic) properties:

Layer	Ω	α_∞	R (Nm^{-4}s)	Λ (mm)	Λ' (mm)
1			(not applicable)		
2	0.97	1.035	9000.	15.	20.

TABLE 3:   **Data for 3-d layered example**

Mechanical properties:

Layer	thickness (mm)	ρ_s (kg/m3)	E (kN/m2)	ν
1	0.8	7820.	206.8x10^6	0.3
2	50.	1305.	134. + 23.5j	0.35 + 0.1j
3	0.8	1626.	13660. + 3220j	0.35 + 0.1j

Porous (acoustic) properties:

Layer	Ω	α_∞	R (Nm^{-4}s)	Q (N/m2)	ρ_f (kg/m^3)
1			(not applicable)		
2	0.96	1.2	5500.	147 400.	1.293
3			(not applicable)		

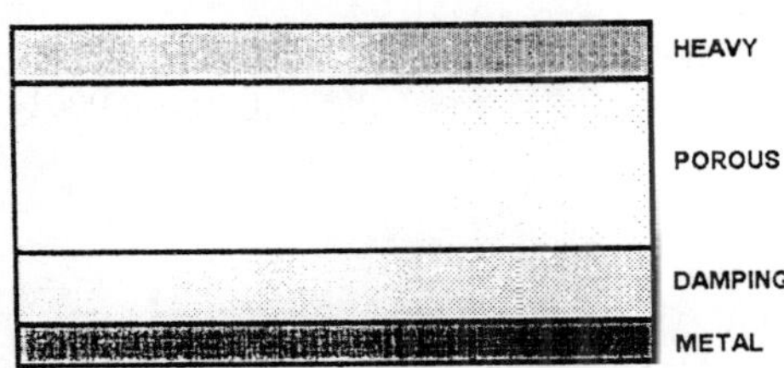

FIGURE 1:   Schematic of multi-layered material

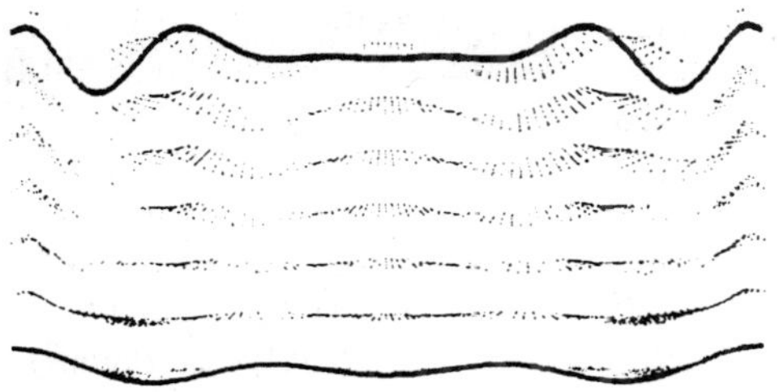

FIGURE 2:   Out-of-phase behaviour between layers

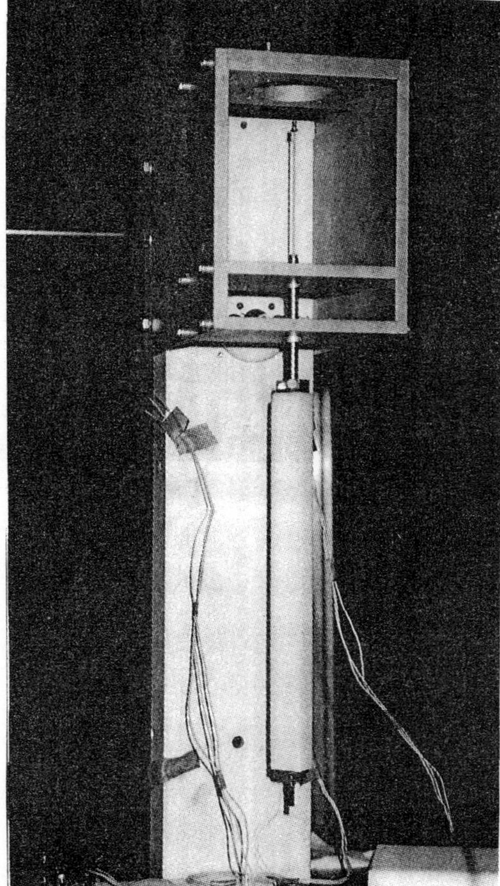

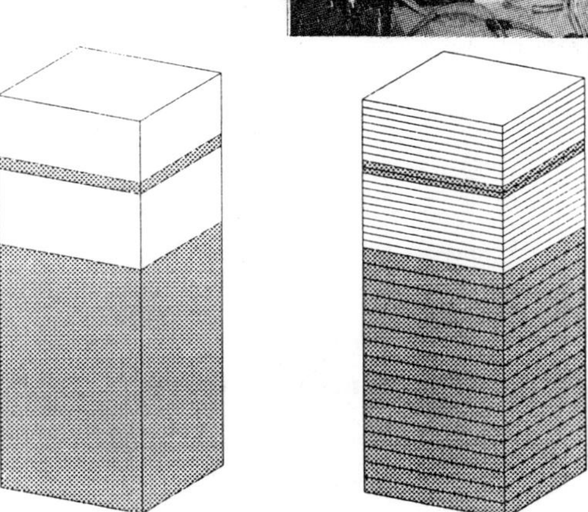

FIGURE 4:   Coarse and fine meshes for surface impedance calculation

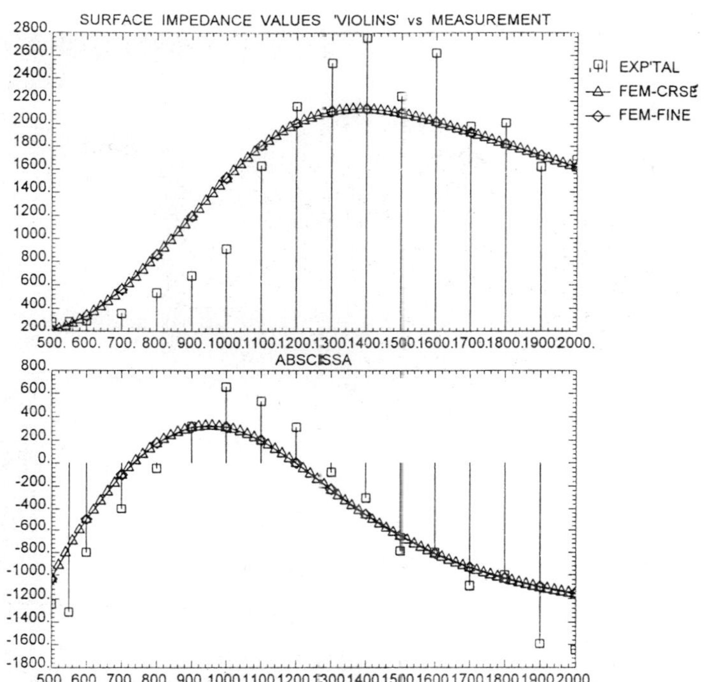

FIGURE 5: Predicted surface impedance (line) compared to measurements (bars)

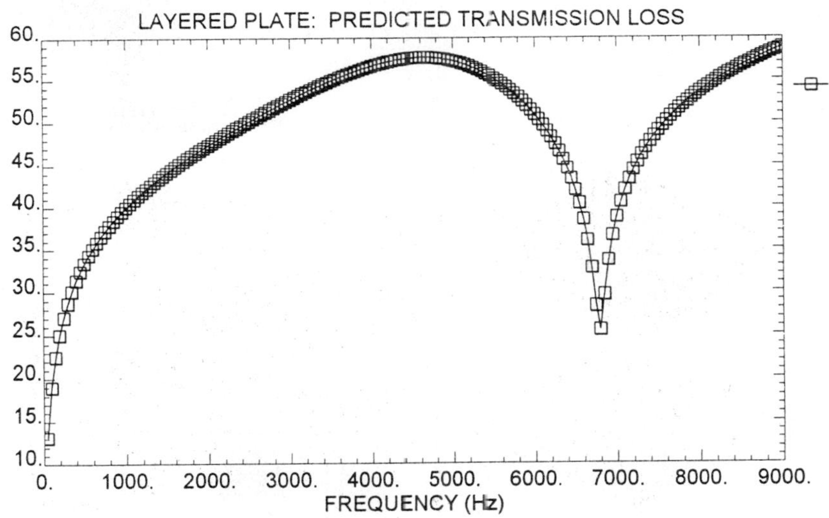

FIGURE 6: Through-panel transmission loss curve (dB)

FIGURE 7: 3-d layered plate: (a) mesh; (b) vibration shape at 100 Hz (c) interior pressures at 300 Hz

(a)

(b)

(c)

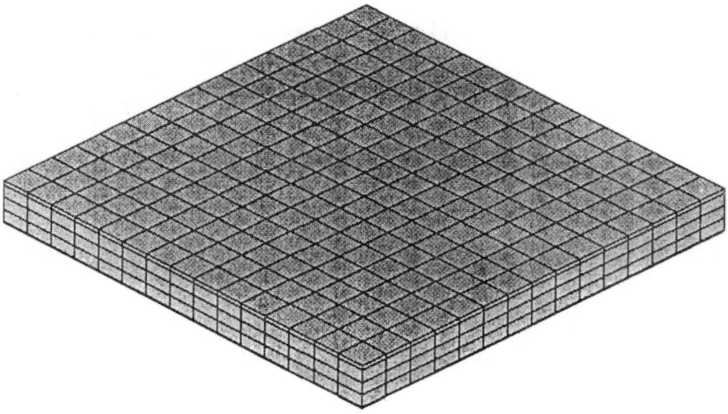

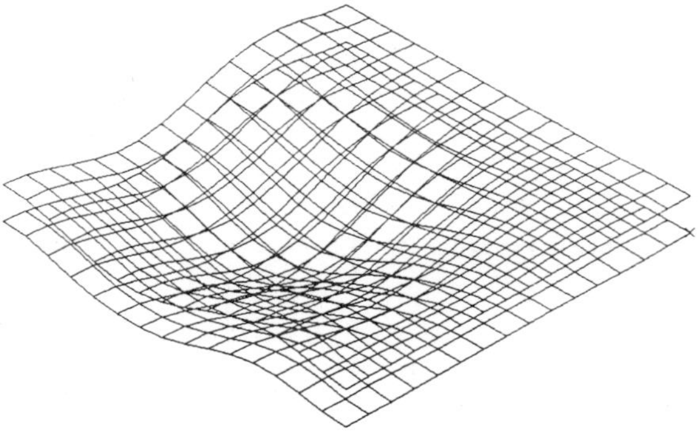

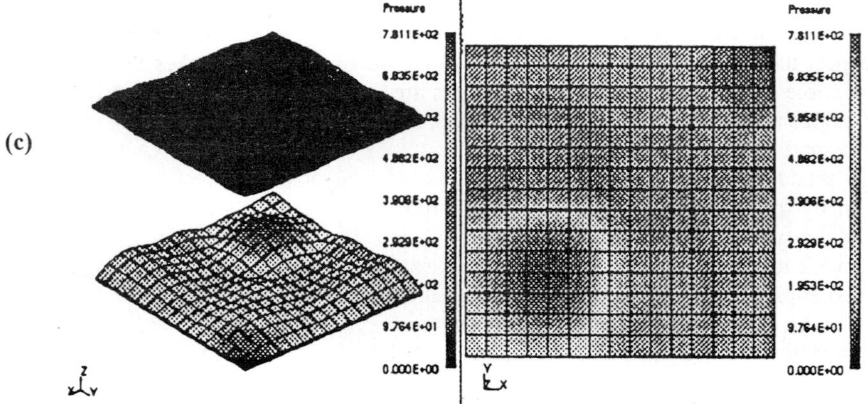

wo-dimensional response of vehicle power units to excitation from inboard constant velocity joints

A LLOYD MA, CEng, MIMechE, A TUCKER-PEAKE BSc, CEng, and S C BARTLETT BSc, MSc
GKN Technology Limited, UK

SYNOPSIS

In front-wheel-drive cars, lateral vibration of the power unit in response to forces from the inboard constant-velocity joints can be a problem ('shudder'). However, in certain vehicles, unpleasant rocking vibration modes are found, indicating a need for a better compromise between the power unit mounting arrangement and the driveline characteristics.

Asymmetric combinations of inboard joints have been shown to influence shudder, as have 'plunging tripode' joints with modifications predicted to alter the phasing and amplitude of the tri-axial support forces and moments. Therefore, a design methodology for overall cost/NVH optimisation of the power unit mounting and driveshaft system can be envisaged.

1 INTRODUCTION

Most cars today have front-wheel-drive. In spite of the mechanical difficulty of transmitting the engine power to the wheels that steer, this is an attractive configuration, because it provides more useable space and needs less machinery than rear-wheel-drive. However, the transmission system depends on the availability of constant-velocity universal joints (CVJ's) of the right performance, durability, and price.

GKN Automotive is the world's leading supplier of these joints, and with its partners, delivered 56 million in 1994. This core product area is supported by major R&D expenditure, part of which is focused on noise-vibration-harshness (NVH) characteristics. Drivers expect their cars to be ever more refined, yet weight reduction makes the vehicles more susceptible to unwanted excitations. Matters are not helped by the larger driveshaft operating angles needed for optimum power unit positioning.

A driveshaft can act as a noise transmission path, and as an excitor (1). One resultant problem is 'shudder', a low-speed vibration of the power unit, sometimes induced under hard acceleration by the inboard CVJ's next to the gearbox. This can be a major issue in the selection of the type of CVJ to match the power unit mounting and suspension layout.

Most work to date has concentrated on the transverse shudder mode. However, more sophisticated engine mounting arrangements have increased the significance of vertical and rocking motions. To progress towards an understanding of this generally 2-dimensional problem, experimental and mathematical modelling techniques, developed earlier (2), have been applied to asymmetric driveline arrangements, and to inboard joints with modified tri-axial excitation patterns.

2 INBOARD (PLUNGING) CV JOINT TYPES

2.1 General

In the early days of front-wheel-drive, it was common for the sideshafts to have a slip-spline, sometimes incorporating ball bearings. The inboard joint then had fixed centres, typically a Hookes joint with rubber bush bearings. Even at low angles, however, the cyclic variation of rotational speed could cause torsional vibrations. Several varieties of true constant-velocity joint were developed that not only addressed that problem, but also accommodated the 'plunge' motion, eliminating the slip-spline (3).

2.2 Plunging ball CV joints

Two types of ball joint are common, both giving acceptable NVH characteristics and heat generation up to an installed angle of about 7°. The VL (*Verschiebegelenk Löbro*) has straight ball grooves in the inner and outer members with alternating helix angles. The 'cross-track' effect, supplemented by a cage, keeps the balls in a single plane (Figure 1). It is a simple joint, with relatively small plunge capacity and high plunge resistance.

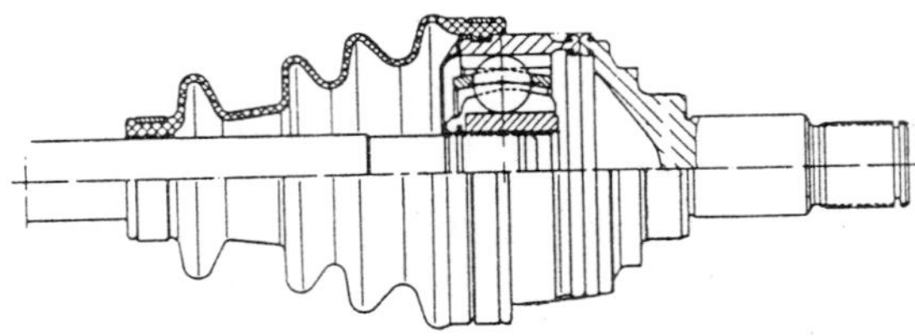

Figure 1 VL joint

The DOJ (*Double-Offset Joint*) also has straight tracks, but they are parallel to the member axes, and the balls are guided by a cage with axially-spaced internal and external spherical locating seats (Figure 2). High plunge capacity is available.

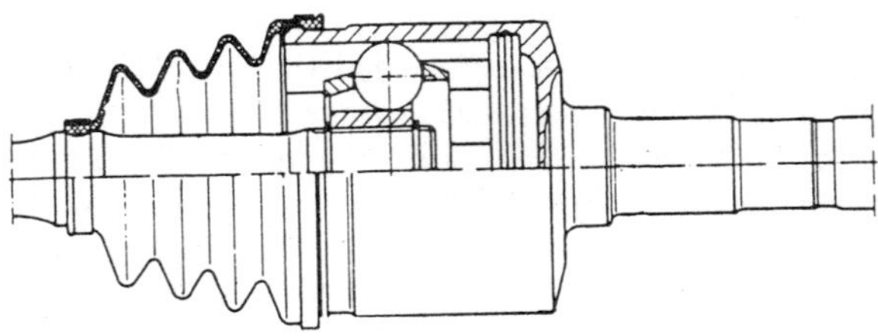

Figure 2 DO joint

2.3 Plunging tripode joints

This type of joint has 3 rollers mounted on a spider. These roll in grooves in the outer member, conferring high efficiency and low plunge resistance (Figure 3).

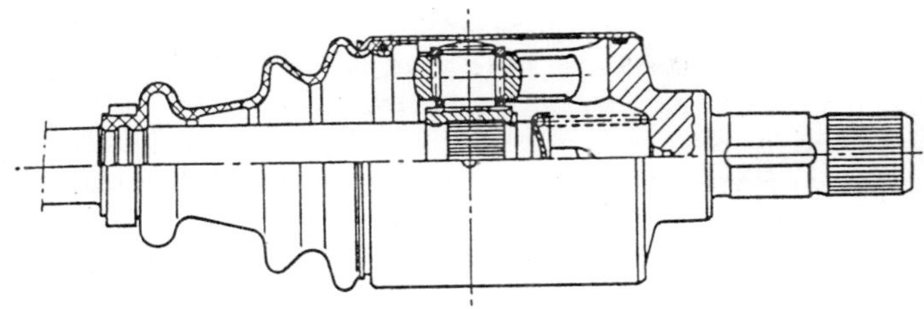

Figure 3   GI joint

The standard GI (*Glaenzer Intérieur*) joint generates 3 axial force cycles per revolution, increasing in magnitude with articulation angle. These are reduced in the GIA, with inclined roller trunnions (Figure 10).

2.4 Compound roller tripode joints

Periodic axial forces may be almost eliminated by tripode joints with compound roller arrangements. These allow the rollers to move in the track grooves with less skidding, and are exemplified by the AAR (*Angular Adjusted Roller*) joint (Figure 4). Sustained operation is possible at angles of 12° or more, without adverse vibration.

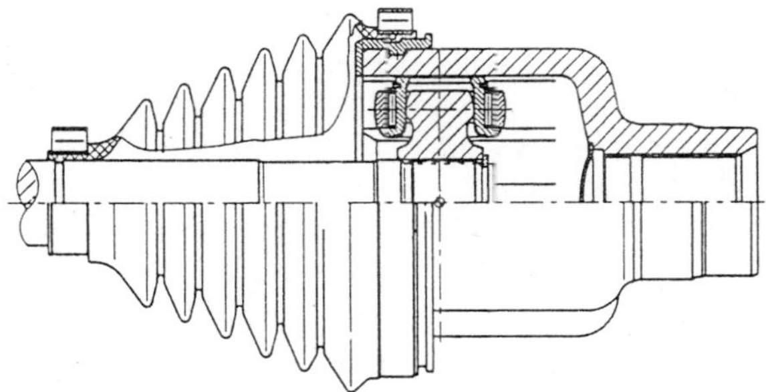

Figure 4   AAR joint

2.5 Driveline asymmetry

Most front-wheel-drive cars have asymmetric drivelines. The sideshafts can be of different length, and sometimes a dynamic vibration absorber is fitted on the longer shaft. However, it is not unknown for the inboard joints to be of different types, since they experience different articulation angles and plunge displacements. This has an affect on various NVH phenomena.

3 POWER UNIT VIBRATION

3.1 Power unit mountings

The power unit mountings in a front-wheel-drive car have to satisfy several conflicting requirements. They have to be soft enough to isolate high-frequency engine firing vibrations, yet stiff enough to react the drive torque in 1st gear, including the torque multiplication of the final drive. To complicate matters, the masses of the power unit components are distributed asymmetrically.

One result of the design compromise is a power unit that has low frequency vibration modes in a plane generally transverse to the vehicle.

3.2 Shudder

Disturbances from the inboard CVJ's easily produce a transverse power unit response, at frequencies up to 20 Hz, which is felt as low frequency shudder. When the relative engine/body motion reaches an amplitude of 1 - 2 mm, the vibration becomes uncomfortable for the vehicle occupants.

Shudder vibration measurements are best made on a rolling road, where the driveshaft torque and articulation angle can be easily adjusted, and the relative phase of the excitations from the two inboard joints can be set for the 'worst case'. Extreme transient conditions that occur during vehicle acceleration may be reproduced and sustained, with suspension height the control parameter for the articulation angle.

Having the vehicle stationary on rollers also permits detailed study of the vibration modes. Running mode analysis provides a full 3-dimensional visualisation of the vehicle and power unit vibration over the speed range of the vehicle. 'Roving' tri-axial accelerometers are used to gather data at specified measurement points.

3.3 Shudder examples

Operating deflection shapes from two different front-wheel-drive cars are shown in Figure 5. In the first example (Figure 5A), periodic axial forces from the inboard CVJ's excite the power unit, producing transverse resonance by the time a speed of 20 km/h is reached.

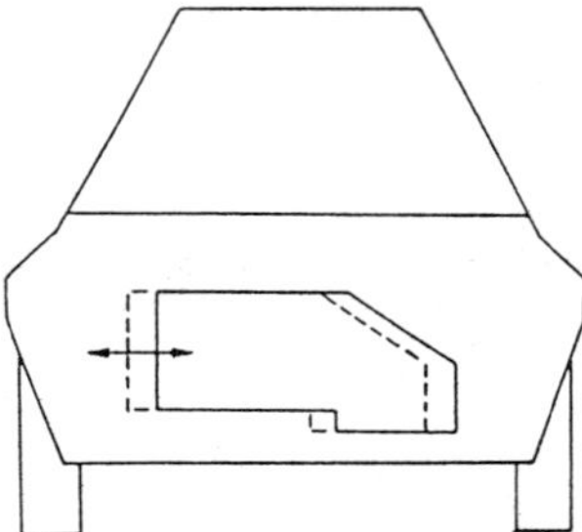

Figure 5A Operating deflection shape for first example car

The amplitude increases with the angle of the inboard (tripode) joint, becoming uncomfortable for the occupants at 8°.

With the second example vehicle, two different operating deflection shapes are evident (Figure 5B).

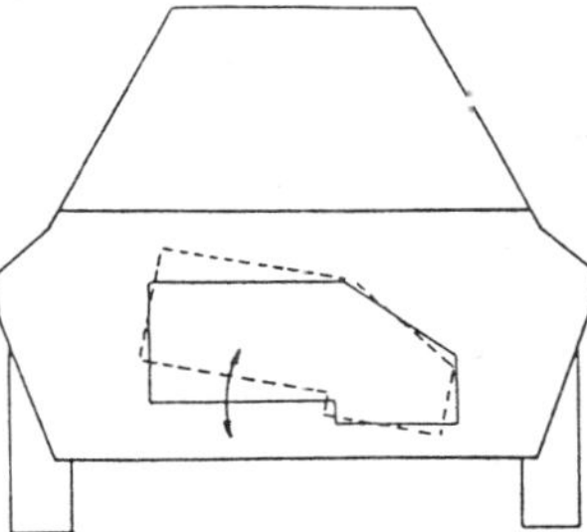

Figure 5B Operating deflection shape for second example car

At 25 km/h, the power unit experiences rocking about an axis parallel to the vehicle centreline. This response then changes to transverse vibration at 31 km/h.

4 EXCITATION FROM THE INBOARD CV JOINTS

4.1 Support reactions

Constant-velocity joints are complex mechanisms. In order to transmit torque through an angle, there has to be an equilibrating support moment, called the *secondary moment*, in the plane of the two shafts, which is balanced by out-of-plane forces acting on the inner and outer members. Forces in the plane of the two shafts are balanced by the perpendicular *tertiary moment* (Figure 6).

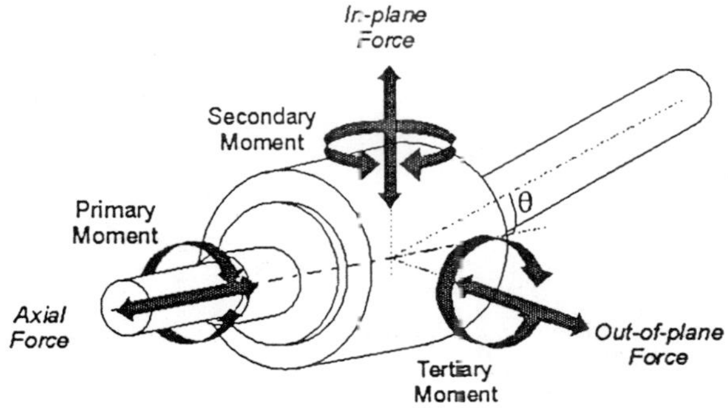

Fig 6 Supporting forces and moments on a constant velocity joint

Furthermore, movement of the internal load-carrying parts, as the joint rotates, causes the internal forces to change their direction, point of action, and magnitude. Frictional forces are superimposed at each contact. In consequence, there is cyclic variation in the forces and moments exerted by the joint members on the supporting structures, namely the power unit and wheel hub. Torsional excitation of the transmission system may also occur.

Excitations from a tripode joint are generally 3rd-order (3 times per revolution), with a 6th-order content in the tertiary moment. Ball joints produce 1st- and 6th-order effects. All the excitations vary with torque and joint angle.

4.2 Effect of periodic axial force and tertiary moment

The effect of axial force on the transverse shudder response is well understood. Objective measurements of this mode have been made both on sensitive vehicles and test benches. New types of joint, such as the AAR and GIA (see section 2) have been developed specifically to reduce axial excitations. The shudder improvement achieved by replacing the standard GI joints with AAR joints on the first example car is shown in (Figure 7). Particular improvement can be seen at high angles.

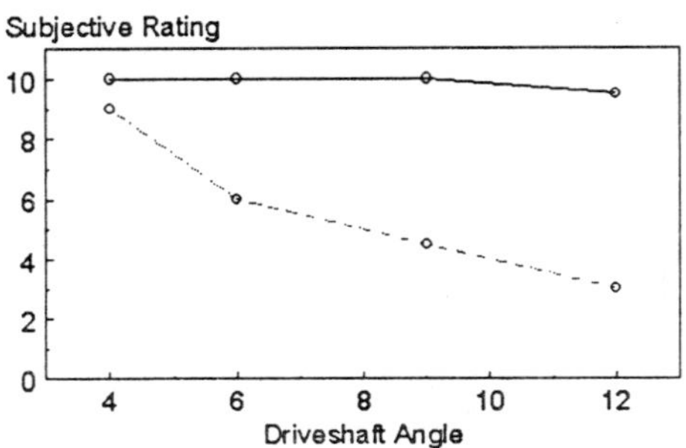

Figure 7   GI/AAR shudder comparison - first example car

With the second example car, however, the reduction in shudder amplitude was much less (Figure 8), and not commensurate with the axial force improvement measured on a test bench.

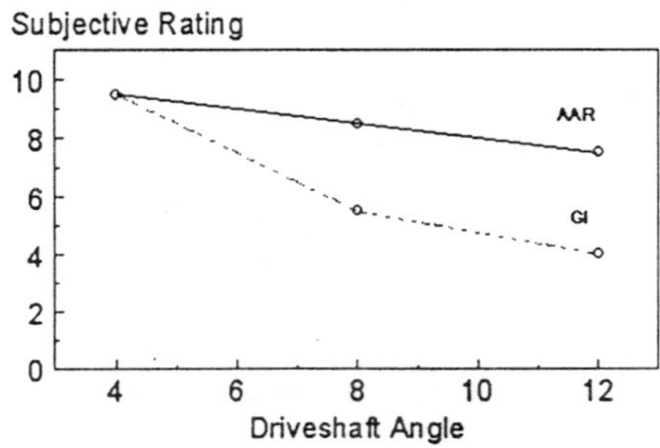

Figure 8 GI/AAR shudder comparison - second example car

Since it has already been established that, for this car, rocking of the power unit is significant, it is fair to conclude that the tertiary moments of the inboard joints, acting generally parallel to the observed rocking axis, are having an effect.

4.3 Prediction and modification of excitations

A mathematical model of a tripode-type CVJ has been developed, taking account of various frictional effects and the detailed geometry of the joint (2). Incremental motion of the joint model produces time domain predictions of forces and moments. Incorporation of further features in the model continues, while work has started on a similar model of ball-type joints. Extensive use of the tripode joint model showed that relatively minor modifications to the basic geometry of the joint had a marked effect on the excitation forces. Reductions resulted mainly from the alteration of the phasing of the internal forces: the frictional power loss did not decrease.

Slight inclination of the tripode trunnions gave axial force reductions of typically 50% (Figure 9), but increased tertiary moment.

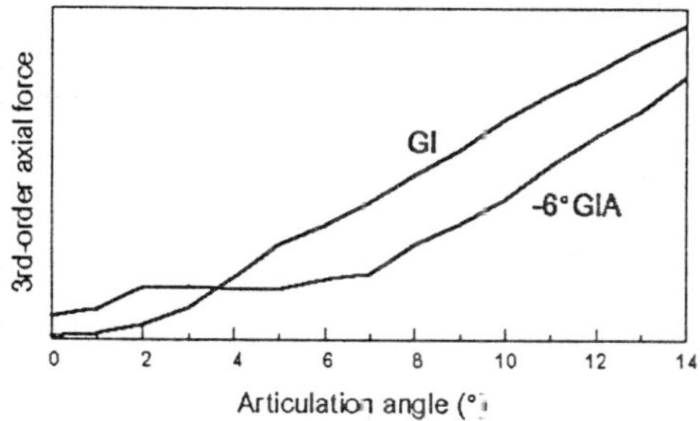

Figure 9 Measured axial force - GI and GIA joints

This modification of the GI joint has been styled the GIA (Figure 10). Both positive and negative inclinations are possible, denoted by GI+A and GI-A, and the predicted results for both are similar.

Figure 10 Prototype GIA tripode assembly

Lateral offset of the tripode trunnions in the plane of the tripode have a significant effect on tertiary moment, and a small effect on axial force. This joint is called the GIO (Figure 11) (4), and, again, positive and negative offset versions are possible.

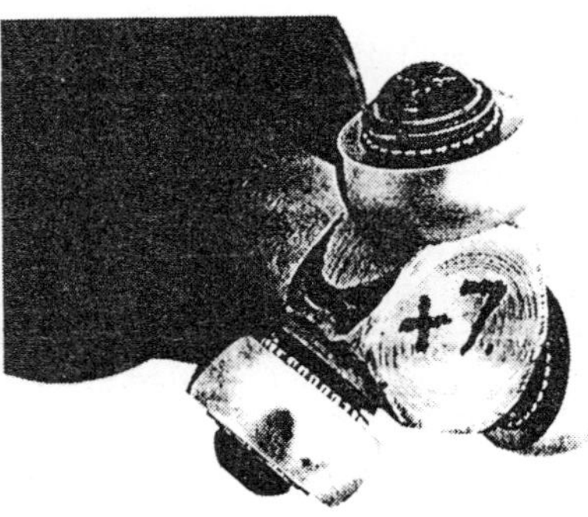

Figure 11   Prototype GIO tripode assembly

The GI-O version is predicted to reduce the amplitude of the tertiary moment in a left-hand driveshaft, and the GI+O to increase it.

Permutations of these modifications are obtained by combining trunnion inclination and offset, to produce the GI+A+O, GI+A-O, GI-A+O, and GI-A-O. All of these are predicted to generate high tertiary moments, and to have a periodic axial force worse than the GIA.

5   PREDICTION OF ROCKING EXCITATION ON POWER UNIT

5.1   Dynamics in the transverse vehicle plane

In the plane transverse to the vehicle centreline, three excitations from each inboard joint affect the equilibrium of the power unit: the axial force, the in-plane force, and the tertiary moment (Figure 6). In this plane, steady-state motion of the power unit at the forcing frequency can be characterised as an oscillation about an instantaneous centre. (For pure transverse vibration, the centre would be an infinite distance above the power unit; for pure rocking, it would be at the centre of mass.)

The response of the power unit to the three excitations from an inboard joint is proportional to their combined moments about the instantaneous centre, calculated in the time domain to take account of the relative phasing. One of the boundary conditions in the model of the (tripode) inboard joint is that there is no transverse moment imposed by the CVJ at the outboard end of the driveshaft. As a result, the centre of the outboard joint is a node: the exciting moment for a mode centred on the outboard joint is always zero.

5.2   Survey of theoretical excitations

The combined 3rd-order moment excitations were calculated about each point on a 100 mm grid in the transverse vehicle plane, for an articulation angle of 11.5°. For all variations of the standard GI tripode joint, there were instantaneous axis positions about which low excitations were predicted. These positions fall on a straight 'null line', through the outboard joint centre, rising in the inboard direction. The excitation moment for instantaneous axis positions off the null line also changed relative to the GI, as shown in table 1.

Table 1 Combined 3rd-order moment excitation characteristics

joint type	null line gradient	off-line change
GI	1:5	datum
GI+A	1:8	-50%
GI-A	1:3	-50%
GI+O	1:5	+12%
GI-O:	1:5	-12%
GI+A+O:	1:12	-40%
GI+A-O:	1:10	-45%
GI-A+O:	1:3	-40%
GI-A-O:	1:3	-40%

5.3 Discussion of model results

Results for the GIA configurations have been experimentally verified with respect to periodic axial force (2), but equivalent work has not yet been done for response involving the in-plane force or tertiary moment. The accuracy of the predictions depends on the correct estimation of each friction factor within the model. Nevertheless, it appears that (a) it should be possible to select an optimal variant of the GI joint for each location of the instantaneous axis and (b) potentially useful behaviour is exhibited by the GI+A, the GI-A, the GI-O, and the GI+A+O.

6 MEASURED EFFECT ON SHUDDER OF ALTERED ROCKING EXCITATION

6.1 Test programme

An experimental programme was carried out with the second example car, for which there had not been a good correlation between the level of periodic axial force excitation and the perceived shudder performance. The objective was to investigate the effect of changing the level of rocking excitation. No consideration was given to other NVH problems, such as 'idle boom', which can be caused by the high plunge resistance in some types of CVJ. Various combinations of joints were tested on the left and right sides of the gearbox, and the shudder performance was rated, subjectively, on a scale of 1 to 10. The joints included the GI, VL, AAR, GI-A, and GI±O.

<u>**6.2 Vehicle test results**</u>

Example rating improvements at 8° articulation angle are shown in Table 2.

Table 2 Shudder rating improvements at 8° articulation angle

left CVJ	right CVJ	shudder rating improvement
GI	GI	datum
VL	GI	+1
VL	VL	+4
GI	VL	+1½
AAR	GIA	+1½
VL	GIA	+2½
GIA	VL	+4
GI-O	GI+O	-2
GI+O	GI-O	+1
GI-O	GI-O	+1
GI+O	GI+O	-½

<u>**6.3 Discussion of vehicle test results**</u>

This vehicle has symmetrical driveshaft angles, but the mass of the power unit is offset to one side. It is noteworthy that mirror-image arrangements of the VL-GI and VL-GIA combinations gave different results, even though the sum of the periodic axial forces were the same. A shudder contribution from the in-plane force or tertiary moment is the likely reason.

7 DISCUSSION AND CONCLUSIONS

There is clear evidence, from the combined experimental and modelling activities, that the shudder response of certain vehicles (second example car) depends on factors other than simply the magnitude of the periodic axial force from the inboard CVJ's. Response modes of the power unit that include transverse rocking may be modified by asymmetric inboard joint combinations and by the use of joints expected to have changed excitation moments in that plane.

If this observation could be converted into a design methodology, overall cost/NVH optimisation of the power unit mounting and driveshaft system could be attempted. Further progress will not be easy, dependent as it is on refining and validating models of both the CVJ's and the vehicle. At GKN Automotive, we have the expertise, test facilities, and innovative design solutions to support such work.

8 ACKNOWLEDGEMENT

The author gratefully acknowledges the permission of GKN Automotive Driveline Division to publish this paper, and wishes to thank colleagues who contributed to the activities reported.

9 REFERENCES

(1) BARON, E., NVH phenomena in constant-velocity joints - a 3-fold approach, xxiv FISITA Conference, London, 1992, 51 - 60, (IMechE C389/277)
(2) LLOYD, R.A., BARTLETT, S.C., and TUCKER-PEAKE, A., Improvements to the NVH performance of tripode constant-velocity joints through mathematical modelling, Conference on Vehicle NVH and Refinement, Birmingham, 1994, 159 - 166 (IMechE C487/029)
(3) SCHMELZ, F., SEHERR-THOSS, H.-C., and AUCKTOR, E., Universal joints and driveshafts, Springer-Verlag, Berlin, Ist English ed., 1992
(4) LLOYD, R., and BARTLETT, S., In-plane inclined tripode legs, International Patent Application PCT/GB95/00334, 1995.

Improvement of low frequency vibration of front wheel drive cars

D J LAIGHT BSc, M PHILIPS BSc, R GENWAY-HAYDEN BSc, D R C CRANG BSc,
and D ADAMS BSc, PhD
GKN Technology Limited, UK

SYNOPSIS

Vehicle refinement levels formerly associated with larger luxury vehicles are now being applied to all cars. At the same time marketing pressure on manufacturers and suppliers is for best in class refinement. Three low frequency vibration problems that detract from perceived refinement in all sizes of front-wheel drive car are discussed. *Shudder* is a lateral vibration and *shunt* is an unpleasant fore-aft shuffle of the whole vehicle. *Idle shake/boom* is a noise and vibration disturbance in automatic transmission FWD cars introduced by changing from "neutral" to "drive" while holding the vehicle stationary on the footbrake.

Human perception of these phenomena depends both upon the exciting forces/moments and the vehicle natural modes of vibration. This paper describes the experimental and theoretical modelling methods required to achieve best in class refinement.

1 INTRODUCTION

Improvement to vehicle refinement is now a high priority in vehicle design. Low frequency vibration phenomena can influence vehicle drivers in their preferred choice of car and are not simple to eliminate. Hard won reductions in new vehicle development time can be dissipated when they occur.

Three examples, *shunt, shudder* and *idle boom*, found in front-wheel drive (FWD) passenger cars, are reviewed. In each case the sideshaft plays a significant rôle in the vibration as either an exciter or passively as an element in a modal system. Sideshafts comprise a bar or tubular shaft connecting two CV-joints.

At the wheel end, adjacent to the front hub, the "outboard" CV-joint has a fixed centre and articulates to more than 45° to accommodate front wheel steer angles. Torque is transmitted by six internal balls and this CV-joint is thus called a *ball joint*. True constant velocity can be maintained using ball joints.

The "inboard" CV-joint, adjacent to the final drive, articulates only to the angle necessary to accommodate suspension travel, typically between 10° and 15°. However, it also plunges to accommodate shaft axial motion. It therefore has no fixed centre. Several designs compete as ideal inboard CV-joints, there being a trade-off between performance and cost.

In this paper only the tripode, whose motion is described in Reference 3, and the DO-joint, a 6-balled joint not dissimilar in principle to the outboard ball joint, are referred to.

2 SHUNT

Shunt, also known as *shuffle* or *tip-in/back-out*, is the vehicle fore-aft rocking motion that can follow a step change in driveline torque. This "step-input function" may be introduced by a sudden throttle change or clutch engagement. It causes a torsional response of the driveline and frequently movement of the power-unit on its mounts. Vehicle occupants perceive shunt as whole body fore-aft motion, as shown in Figure 1 below:

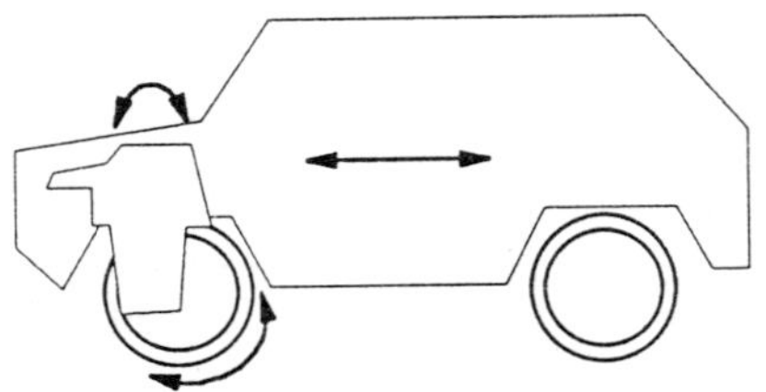

Fig 1 Shunt Reactions in a FWD car

After the step-input function is applied, the vehicle body oscillates due to its inertial mass being coupled to driveline torsional stiffness, as well as to ground via bushings, suspension and tyres. Power-unit pitch motion can also play a significant part (Reference 1) and is included in the half-car model proposal of Figure 2. This model would include all linear and non-linear stiffnesses as well as the appropriate masses and inertias.

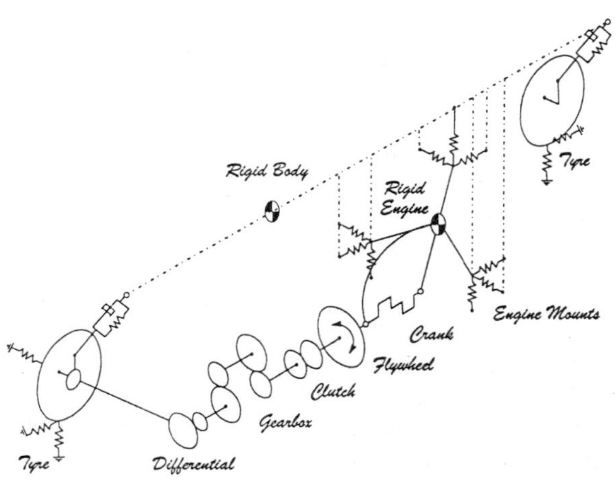

Fig 2 Half-car Model of Vehicle Shunt

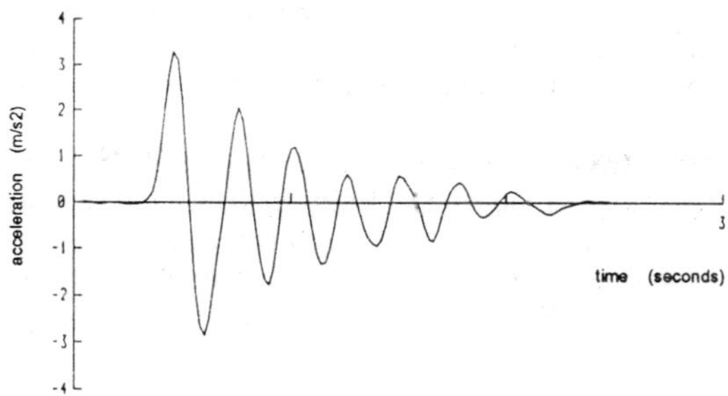

Fig 3 Time History of Chassis Fore-aft Shunt Response

Figure 3 shows a measurement result of longitudinal vehicle body acceleration, which resembles the decay of a damped single degree of freedom system. It is clear that only the first shunt mode, with a frequency in the range 1-5Hz is significant. Since data for the half-car model is sometimes difficult to obtain, a simplified model, neglecting the movement of the power-unit on its mounts, has been employed to date. It is ideal for studying the driveline influence on the chassis fore-aft shunt. Ignoring the power-unit motion is justified in Reference 2. The simplified model is shown in Figure 4 below:

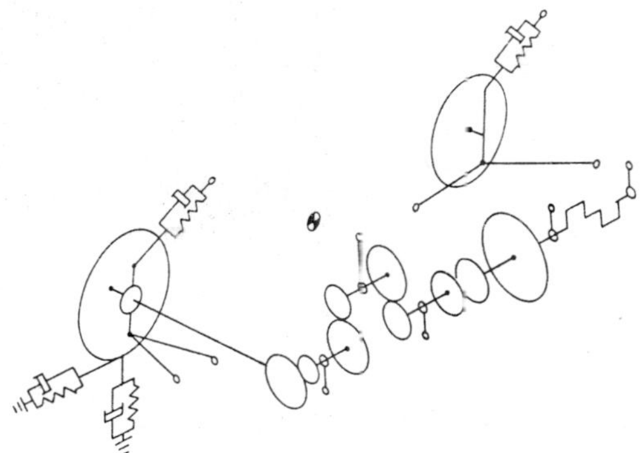

Fig 4 Simplified Model without Power-Unit Mounts

This simplified model can be solved for natural frequencies and modes of vibration as well as forced responses from step-input functions. In the cases tested, time-step integration solutions closely reproduced the test results of Figure 3.

Sideshaft torsional damping measurements obtained from a purpose-built test rig were employed in this model. Changes in shunt performance were insignificant within the range of values measured. Sideshaft torsional damping proved to be negligible compared to other sources of damping, such as the clutch and tyre.

Passenger comfort can be improved by modifying either the level or the duration of the system response to the step-input function. Example parameter changes are:

- add driver-input control measures to prevent step-inputs
- reduce driveline backlash and increase torsional stiffness, to raise the natural frequency of the shunt oscillation and thus limit the duration of each event
- add significant torsional damping into the driveline to attenuate peak shunt response

An automatic transmission, for example, gives a significant increase in driveline system torsional damping. It reduces the susceptibility to driver-induced shunt compared to the equivalent vehicle with manual transmission. The above measures not only reduce shunt, but also the peak sideshaft torque, which itself influences sideshaft design. With the aid of a driveline model the various control strategies can be quickly compared.

3 SHUDDER

Low speed shudder is the unpleasant vibration felt under acceleration within the vehicle speed range 15-30km/h. FWD vehicles fitted with tripode plunging CV-joints are most susceptible. Tripode excitation forces, called *periodic axial forces*, act along the sideshaft axes as illustrated in Figure 5. Their vibration period is three times per sideshaft revolution, on account of the three internal pegs of the tripode design, and is called *third order*. Power-unit resonance on its elastomeric mounts in the frequency range 8-17Hz is the primary response.

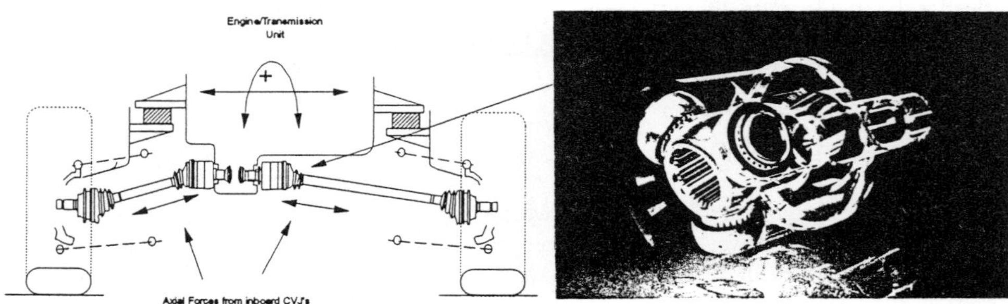

Fig 5 Inboard CV-joint Shudder Forces

The oscillating motion of the power-unit is in turn reacted by rigid-body motion of the vehicle body. Factors found to affect shudder are:

- high torque and maximum inboard CV-joint articulation angles: hard acceleration, causing the front of the car to rise, gives the worst case of both
- tripode CV-joint left-to-right rotational phasing: cross-phasing (▲▼ as opposed to ▲▲) is the worst case since one CV-joint "pushes" while the other "pulls". See Figure 8
- rigid body modes of the body and power-unit excited into resonance

Even with a ball type inboard CV-joint, such as the double-offset joint (DO-joint), shudder can be felt in sensitive vehicles. Shudder vibration is now felt at high speed, since the main vibration period of the DO-joint is once per revolution, or *first order*. It is this change of excitation order that moves the range of DO-joint shudder to three times the vehicle speed of tripode shudder. High speed shudder is truly intermittent and only ever occurs when running for long periods on extremely smooth straight highway. In order to provoke it, lubricant has to be wiped away from the DO-joint internal contact points.

Shudder can largely be prevented by addressing the design and maximum articulation angle of the inboard CV-joints. Excitation forces and moments for different types are now well predicted and examples generating very low periodic axial forces have been developed. Mathematical models of the inboard CV-joint are discussed in Reference 3.

Test rigs have been built to measure the forces and moments generated when CV-joints run under torque and articulation angle, as shown in Figure 6. These rigs are free of resonance up to 100Hz and have dynamically stiff end-conditions for the sideshaft. A system of rotating tri-axial force cells measures force and moment time histories. The mathematical CV-joint models have been correlated against test results from these test rigs.

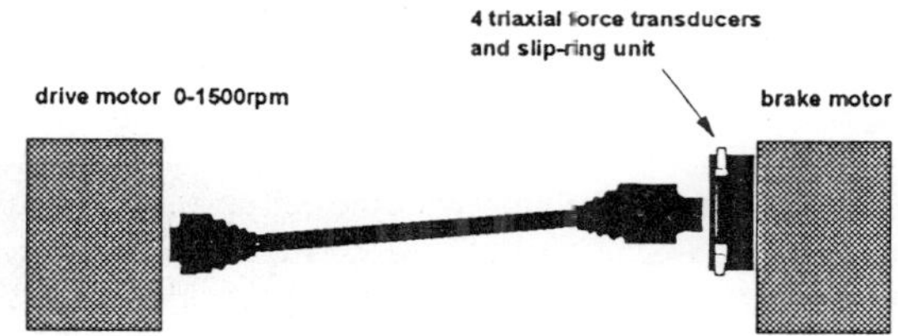

Fig 6 Test Rig for Measurement of Periodic Axial Force

Vehicle sensitivity is represented by (weakly non-linear) frequency response functions showing motion at the driver's seat in different axes per unit applied periodic inboard CV-joint force. The model for sensitivity is shown in Figure 7 below.

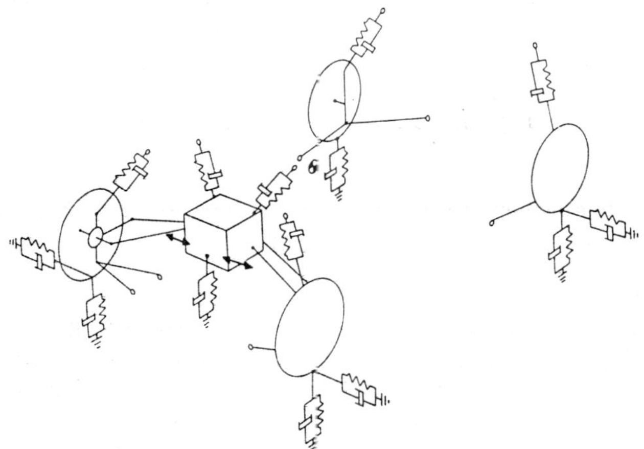

Fig 7 Shudder Sensitivity Model

Discrete components are assembled using the building block approach. Again non-linearity is taken into account when deriving the stiffness modules.

The sensitivity model has been correlated against a standard vehicle dynamometer shudder test in which the vehicle is accelerated slowly in a sweep through the shudder response frequency range under constant torque loading. Test sample sideshafts are each run-in and measured on the test rig described above. The phases of the inboard CV-joints left-to-right, which would normally be random, are controlled, as are the inboard CV-joint articulation angles. Vehicle vibratory response is measured at both the front chassis rail and the driver's seat rail, along with the vehicle occupant subjective rating.

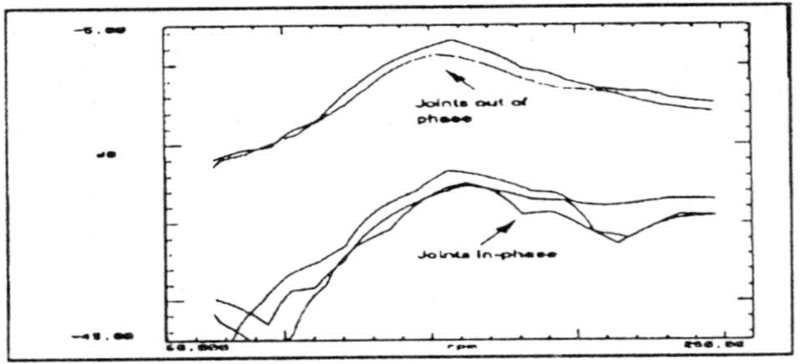

Fig 8 Shudder Test Result from Chassis Dynamometer

Figure 8 shows the effect of driving with the left-to-right tripode CV-joints aligned in phase compared to cross-phased. Cross-phased excitation forces, the worst case periodic axial force functions, are employed in analyses using the shudder model.

4 IDLE BOOM

Idle boom/shake is the power-unit vibration and low frequency noise disturbance at idle. Vibration and sound (noticed as ear pressure) often increase in automatic transmission vehicles during the shift from "neutral" to "drive". Subjective assessments are made with transmission "drive" selected. Rattles, panel modes, steering column, seat and acoustic modes cause tactile and acoustic sensations that are all included in a single rating. The vehicle is driven between ratings to alter the CV-joint internal contact conditions.

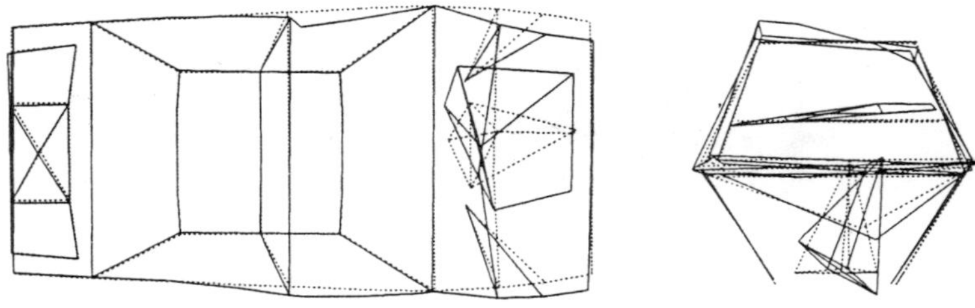

Fig 9 Operating Deflection Shape at 800rpm Idle Speed

Idle boom is excited by the engine combustion cycle and amplified by vehicle body resonances. In this test the sideshafts are brought under torque while remaining stationary. Figure 9 shows a measured vehicle operating deflection shape. Typical engine firing frequencies driving the idle boom are shown in the following table.

Table 1 Engine fundamental firing frequencies at idle

engine type	engine firing order	firing frequency at 760 rpm
3 cylinder	1.5	19.0 Hz
4 cylinder	2.0	25.3 Hz
5 cylinder	2.5	31.7 Hz

Smaller 3-cylinder FWD cars have excitation frequencies which are less audible. Moreover, their body bending resonances, which aggravate the tactile sensation, are not normally excited. Idle vibration/boom in FWD vehicles having transverse engines seems to be less influenced by the CV-joint type than vehicles with longitudinal engines. Therefore experience suggests that larger 4 and 5-cylinder FWD cars, especially those with longitudinal engine layout, carry the greatest risk.
Idle boom is a case study in modal analysis. Elastomeric mounts, exhaust hangers and sideshaft all provide coupling between the power-unit and body. There are two mechanisms by which the inboard CV-joint might influence idle boom behaviour:

•as a constraint to modify power-unit vibratory motion, affecting all transmission paths
•direct transmission of vibrational energy from the power-unit along the sideshaft

Measured power-unit operational deflection shapes altered very little either with or without constraint from high plunge resistance CV-joints. Furthermore, large tuned absorbers located on the front wheel hubs considerably influenced the idle boom level. Therefore the second mechanism is now considered the more significant.

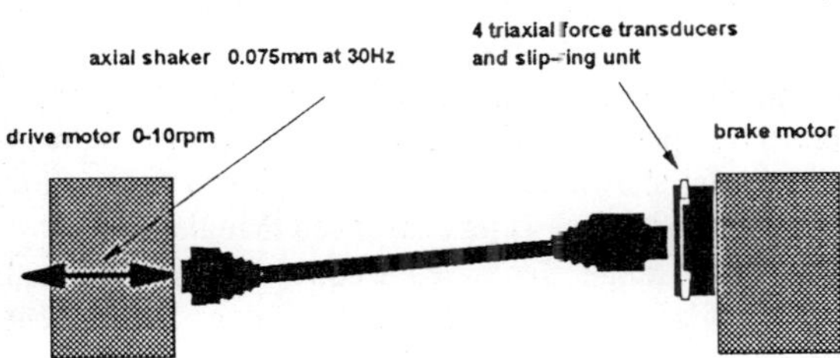

Fig 10 Test Rig for Measurement of Plunge Resistance

The CV-joint influence is resistance to the small plunging oscillations of about 0.075mm amplitude of the idling engine. Friction between the stationary CV-joint internal surfaces, locked together by the torque moment, can vary considerably. Hence each rating assessment is checked several times.

Plunge resistance test rigs, illustrated in Figure 10, have been built. Sinusoidal motion of amplitude typically 0.075mm is applied at 30Hz to the inboard CV-joint held under articulation angle, torque and slow rotation. A slow rotation of the CV-joint brings new surfaces into contact during the test cycle. Again there are no rig resonances below 100Hz and the same load transducing system measures transmitted forces and moments. Results have been used to correlate CV-joint dynamic frictional models.

CV-joints with measured plunge resistance have been assessed in different vehicles. Measurements are made using an acoustic head and several vibration sensors to cover the number of sensations that occur simultaneously. The full vehicle vibratory motion shown in Figure 9, its operating deflection shape, is evaluated using running modes, as described in Reference 4. Idle boom suppression measures are:

- control of engine idle speed to tune its' excitation frequency away from the offending resonances
- low axial plunge resistance inboard CV-joints: cases of a high plunge resistance CV-joint being better do exist, but are extremely rare

A full idle boom model must contain significant characteristics of the vehicle body coupled to engine vibration up to 35Hz, which is not feasible for a component company working alone. Therefore the reduced model of Figure 11 is proposed.

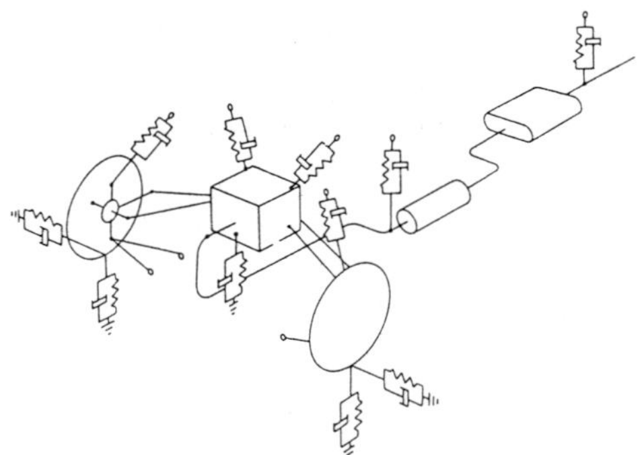

Fig 11 Reduced Model for Idle Boom

This model is limited to estimating the motion of the idling engine and the forces transmitted to body-connection nodes. Since idle boom modelling requires full body-in-white modal data, the reduced model shown above will be employed for the forseeable future.

Progress on idle boom modelling is less advanced than the other two topics discussed, owing to its higher frequency range. Increased frequencies mean that the assumption of lumped mass no longer holds for the vehicle body. Other component interactions are complicated as a result.

5 SUMMARY

In each case of low frequency vibration the fundamental mechanism is well understood. Hypotheses of cause and effect have been verified through driveline hardware modifications that influenced each problem in a predicted manner. The following table summarises the descriptions of the three examples:

Table 2 Summary of low frequency vibration phenomena

phenomenon	excitation	response	frequency range	driveline influence
shunt	driveline step-input function	chassis fore-aft mode	1-5Hz	torsional stiffness, damping, backlash
shudder	inboard CV-joint periodic force	power-unit modes	8-17Hz	inboard CV-joint periodic force
idle boom	engine firing frequency	body bending modes	18-35Hz	inboard CVJ plunge resistance

Even today, driveline refinement engineers rely heavily on vehicle experimentation, the hardware change and re-test method. However it comes too late in the development programme for major design alterations and propagates the risk that a low frequency vibration problem might hold up a vehicle development program. This risk can only be avoided by employing predictive modelling tools. The effort required to complete each correlated model depends upon:

- the availability of accurate data
- the degree of correlation with experiment required in the model
- the solution frequency range, as was pointed out in the discussion of idle boom

In the past, prohibitive costs have also been an argument against the predictive modelling approach advocated, but with computers and software becoming more powerful, the opportunity is there to be grasped.

6 ACKNOWLEDGMENT

The author gratefully acknowledges the permission of GKN Automotive Driveline Division to publish this paper and wishes to thank the colleagues who contributed to the activities reported.

7 REFERENCES

(1)   Rooke GS,Crossley PR, Chan EA (Ford Motor Co Ltd) 1993
        Computer Modelling of a Vehicle Powertrain for Driveability, Autotech 1993
(2)   Manfred Mitschke
        Fahrzeug-Ruckeln, Automobiltechnische Zeitschrift 96 (1994) 1
(3)   Lloyd R, Bartlett S, Tucker-Peake A (GKN Technology) I.Mech.E 1994
        Improvements to the NVH performance of Tripode Constant-velocity Joints through Mathematical Modelling, Vehicle NVH and Refinement, B'ham 1994
(4)   Laight DJ, (GKN Technology), 1992
        A Running Mode Study of Idle Vibration, Symposium on NVH Challenges, Problems and Solutions, Crowthorne, Oct 1992

Source evaluation and modification for conformance with EC drive-by noise legislation in accordance with Directive 92/97/EEC

P S GRIFFITHS BEng
MIRA, UK

SYNOPSIS

With the recent introduction of more stringent drive-by noise legislation in Europe, car manufacturers are having to identify carefully which areas of their vehicles require refinement in order for them to meet the targets. Through an objective approach, the main sources which contribute to drive-by noise can be targeted to develop cost effective solutions.

This paper presents work undertaken at MIRA to evaluate the main sources contributing to the exterior noise of a vehicle during drive-by according to directive 92/97/EEC. A case study is presented showing some detailed results. The paper looks at the shielding technique used to identify the sources, the accuracy of the technique and the factors which influence the results. The paper also examines how the data produced was used following the source ranking to help reduce the main contributing sources of a current production vehicle.

1 INTRODUCTION

Drive-by noise levels are currently the main aspect of passenger vehicle noise emissions which is covered by legislation. This legislation requires a representative vehicle to perform certain tests to produce a resultant noise level which is less than a predetermined maximum.

A recent change to the legislation has introduced more stringent drive-by noise levels for passenger cars. This change in legislation has led vehicle manufacturers to focus more effort on reducing the exterior noise during the drive-by test. Manufacturers are having to identify the major sources throughout the drive-by test and rank them in order of importance. Careful ranking of the sources ensures that time and effort can be directed at the most important areas and an optimum solution can be found with a minimum of time and cost.

MIRA has recently undertaken a project to source rank a current medium sized diesel saloon production car and has worked with the manufacturer to develop solutions so that the vehicle can meet its target of 74 dB(A). This paper examines the technique used and discusses the importance of effective shielding on the quality of results achieved.

2   BACKGROUND

The car under test was a current production vehicle which complies with 92/97/EEC. The drive-by work was part of a model year improvement programme. Scope for development of practical solutions had to be within existing limited constraints of packaging and existing design. Standard drive-by tests, in accordance with the new standard, had been carried out with a sound level meter on four cars. Drive-by figures ranged between 73.9 and 74.8 dB(A).

The proposed technique for evaluating the sources was a shielding method, using an open window technique. It is generally accepted that full treatment of a vehicle is difficult, but to improve the accuracy of the results it was proposed that the test vehicle drive-by level, when fully clad, be reduced by at least 10 dB(A). With a 10 dB difference in level, the combined contribution of two uncorrelated sources would be 0.5 dB higher than the higher source level. Therefore for dominant sources, the potential accuracy would be 0.5 dB. Reducing the noise to such low levels would mean that all sources had to be well attenuated. The technique is well proven and accepted.

3   SOURCE RANKING METHOD

3.1   Technique

Before starting any test work, consideration should be given to the areas that are likely to contribute to drive-by noise. In general, the main areas are: tyres, intake and exhaust orifices, intake and exhaust systems and powertrain. The contribution of each of these areas needed to be deduced. For the case study, the powertrain radiated noise, intake radiated noise and exhaust downpipe were considered to be one item called engine bay. Also the front and rear wheels were considered separately.

In order to carry out processing of the data for source ranking, during each test the following data was acquired: vehicle speed, vehicle position, engine speed and exterior noise on both sides of the vehicle. By taking this data, a full definition of the vehicle path through the drive-by area could be ascertained. To achieve repeatable results, the weather conditions needed to be consistent.

3.2   Noise reduction measures

Cladding of a vehicle must be carefully carried out. When cladding, care must be taken to ensure that each source can be individually exposed without effecting the attenuation of a second source. Cladding should be designed to work across a wide frequency spectrum.

During cladding the weight of the vehicle is increased and can be an important factor during the drive-by test.

The intake and exhaust orifices are best attenuated using large additional silencers. These should be ducted to an outlet at least 4 m above ground in order to move the orifice noise as far from the test area as possible.

Moulded slick tyres of equivalent rolling radius help reduce tyre noise. In addition a skirt around each tyre from an absorbing material will reduce the contribution from each tyre.

The remaining sources can be reduced by a combination of lead quilting and fibreglass. The engine should be enclosed in lead and the rest of the engine bay can be filled with fibreglass. If engine cooling is a problem, then the radiator should be mounted outside of the engine bay. Additional quilt on the outside of the engine will also help.

To reduce any wind noise during the drive-by test, any feature which might cause a problem should be either removed or smoothed out as much as possible.

3.3   Results for case study

The effect of adding weight to the car had a negligible effect on the vehicle performance during the drive-by test. See figure 1.

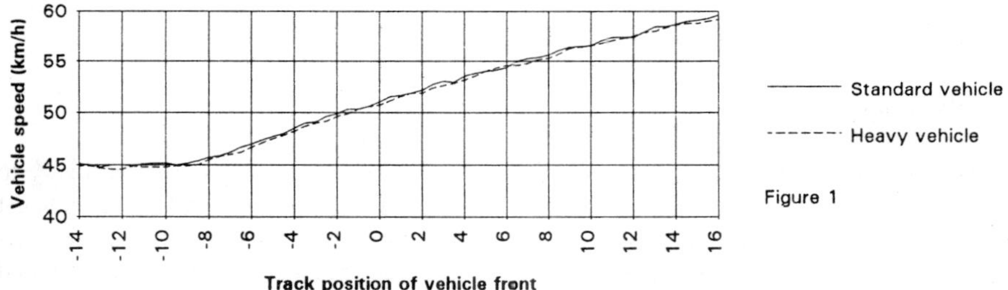

After fully cladding the car, a reduction of 11.3 dB(A) was achieved in the drive-by level.

Figure 2 shows the individual noise source contributions. The figures show a comparison of the overall level compared to the contribution of each source.

Figure 2

2nd Gear LHS
2nd Gear RHS
3rd Gear LHS
3rd Gear RHS
SPL (dBA)
Track position of vehicle front
Standard vehicle
Tailpipe
Front Tyres
Rear Tyres
Exhaust Radiated
Intake Orifice
Engine bay

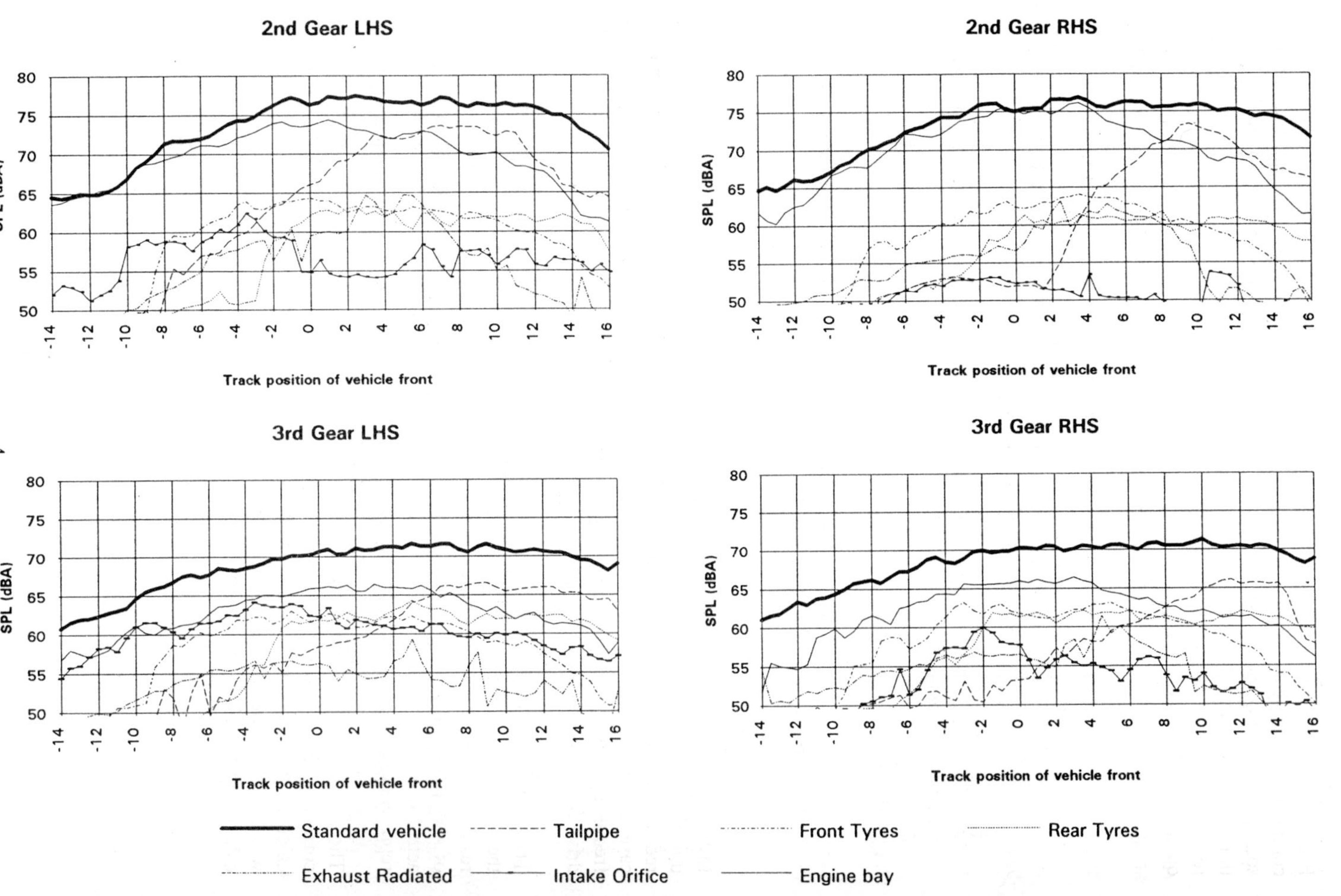

3.4   Discussion

From figure 2, it can be clearly seen that, for second gear, there were two major sources contributing to the drive-by noise. At the beginning of the tests the engine bay was the main source, and during the second half of the test the exhaust tailpipe orifice dominated. For third gear the same two sources had the largest contribution, but their dominance was reduced. In order to reduce the drive-by level, these were the main sources identified for quietening treatment.

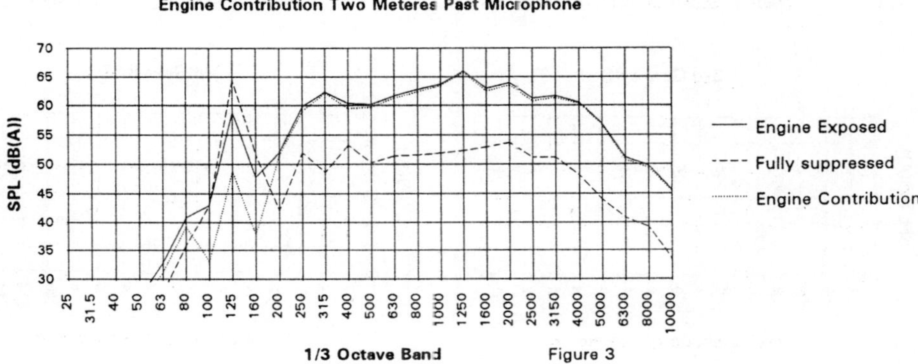

In figure 3, it is shown that the noise reduction due to fully suppressing the vehicle, is not uniform over the frequency range. For the engine contribution, there is a good uniform reduction above 200 Hz. At the firing frequency there is no reduction. In this particular case, suppression of the firing frequency (125 Hz) is not of prime importance, since the high frequency reductions indicated for the engine bay and tailpipe are sufficient to reduce the drive-by noise to within the legal requirements.

The validity of the results obtained was checked by summing the contributions from each of the sources to arrive at a predicted overall level which was compared with the measured result. Figure 4 shows the results of this comparison. Results from the main drive-by area of -10 m to +10 m, show good correlation between the actual and predicted values. The actual drive-by value was the data taken from the standard vehicle. The predicted value represents the sum of the individual sources.

The good correlation shows that the initial estimates of accuracy of dominant sources were correct. This correlation enabled the experimental data to be used with confidence to calculate the reduction required in the dominant sources in order to achieve 74 dB(A). This was done by reducing the contribution of the engine bay and exhaust tailpipe orifice, and calculating the effect on the drive-by figure.

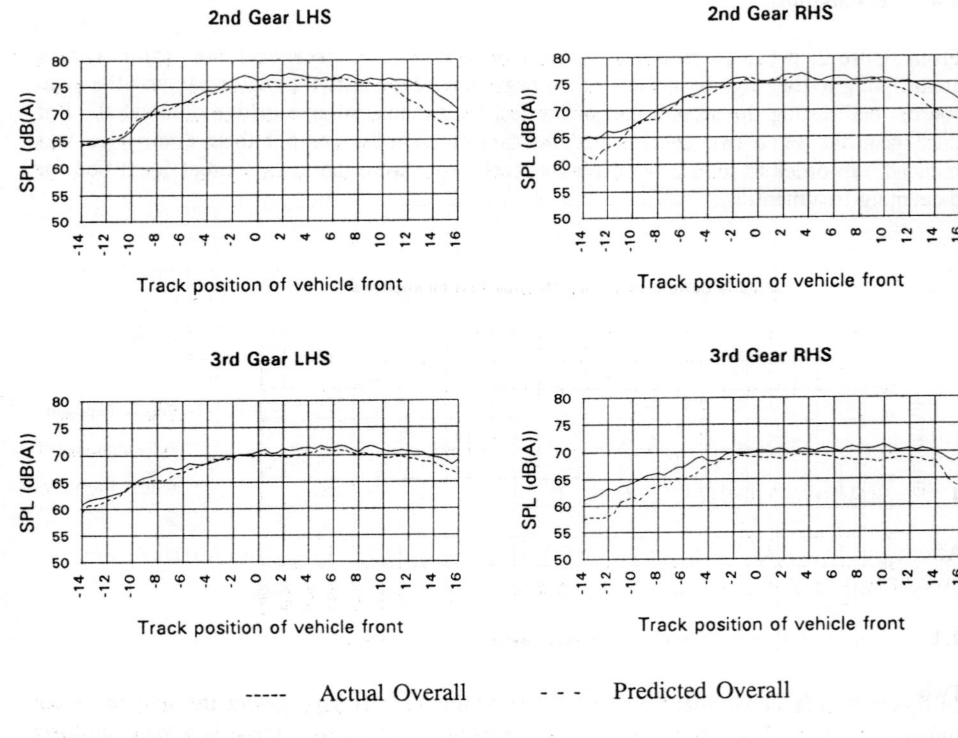

Figure 4

It should be noted, however, in figure 4 that the predictions of the drive-by levels are lower than the measured data. The reason for this is that the contribution from each source is also under predicted by virtue of the calculation procedure which is used to define the individual contributions. Using the windowing technique, the contribution of the individual sources is taken as the difference between levels measured with the source exposed relative to the silenced vehicle. Unless the level achieved by the silenced vehicle is low enough, the contribution of each source will be underestimated, with the error being closely related to the 0.5 dB accuracy figure quoted earlier.

However, by ensuring that the fully clad vehicle achieves overall reductions of over 10 dB(A) then under predictions should be limited to 0.5 dB as indicated by figure 4. This under prediction is also a useful indicator that the open window remains substantially uncontaminated by contribution from the other vehicle sources for a particular window.

Having identified the main sources, the frequency content of these sources was examined. Figure 5 shows the frequency content for the exhaust orifice during 2nd gear on the right hand side of the vehicle. It shows that the 1/3 octave bands from 630 to 2000 Hz were significant. This information was used for system development when attempting to reduce the source contribution.

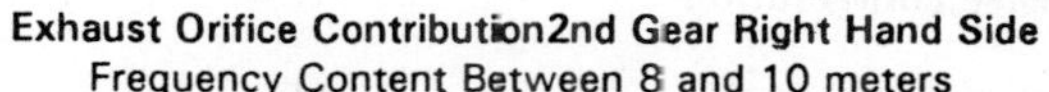

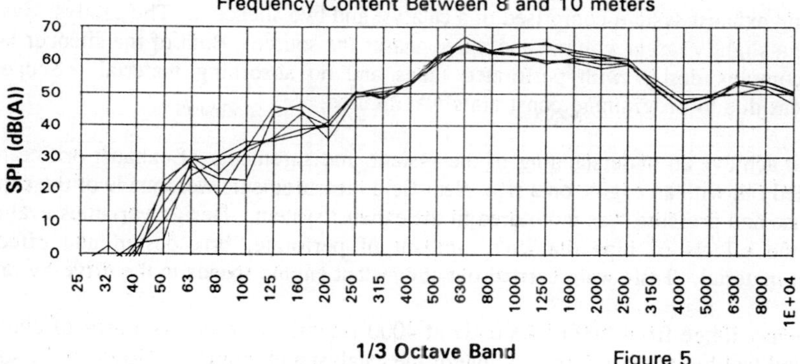

Figure 5

4 SOURCE MODIFICATION

After identifying the two dominant sources contributing to drive-by noise, each of them was investigated in turn.

4.1 Engine bay contribution

Two options were considered in order to reduce the engine bay contribution.

a)   Reduction of noise by treating the source.
b)   Improved encapsulation of the engine bay

After discussion with the vehicle manufacturer, it was agreed the best method to reduce the engine bay contribution was to improve the encapsulation of the engine bay. This was due to the limited scope for changes to the engine itself.

The existing engine bay acoustic treatment comprised an under bonnet liner, an undertray fitted to the subframe with some acoustic absorption, wheel arch liners, and small front side shields mounted from the wheel arches.

A further mini source evaluation exercise was carried out to identify where the engine bay sealing could be improved.

The results indicated that an enlarged undertray together with an additional side shield fitted to each side at the rear of the wheel arches would be required.

4.2   Exhaust orifice contribution

The existing exhaust system comprised of a catalyst and two silencers. The catalyst was due to become a slightly larger volume and move nearer the source. Both of the silencer boxes were of complex design with perforated tubes and no absorbing material. Scope for development due to programme constraints was limited.

In order to achieve an understanding of the system, measurements of exhaust orifice noise were carried out with an engine on a rig. Near field measurements were made of the exhaust orifice noise and pressure was monitored in the exhaust system. Design variables evaluated included the effects of pipe diameter, amount of perforate, box design and effect of absorption material. Tests were carried out at constant engine speeds in the drive-by range.

The frequency range from 500 to 2000 Hz at 4000 rev/min was the main area of concern. It was found that by increasing the volume of the catalyst and moving it closer to the source and introducing absorption into the rear half of the second box, an overall reduction of 9 dB(A) was achieved at the exhaust orifice. This was done without increasing back pressure

5   VERIFICATION

The improved engine bay sealing and proposed exhaust were tested on a vehicle during drive-by. The overall effect was a reduction of 0.9 dB(A) in drive-by.

6   CONCLUSIONS

Application of an open window shielding technique has been shown to produce results of sufficient quality to allow an accurate drive-by source ranking. Accuracy of the ranking is improved by increasing the reduction in noise resulting from shielding. A reduction of 11.3 dB(A) was achieved for the case study.

Required reductions in sub-system contributions were identified from the source ranking, enabling development actions to be identified.

For the subject vehicle of this paper, the main contributors to external noise during the drive-by tests were the engine bay and the exhaust tailpipe orifice. Improvements to the engine bay shielding and modifications to the exhaust resulted in a reduction of 0.9 dB(A) to the total drive-by noise level.

Practical applications of sound quality analysis techniques

R M S MAUNDER BEng
Ricardo Consulting Engineers Limited, UK

SYNOPSIS

A product's sound quality is as important as its look and feel in conveying an image to the customer. A sound quality ideally suited to the product's target market can be selected and shaped using computer-aided analysis techniques such as recordings of vehicle build modifications, digital modifications to these recordings, intake and exhaust noise predictions, subjective jury tests, objective analysis, and predictive indexes.

This paper describes the application of these techniques to a range of vehicle engineering problems (see Figure 1) both at the concept and development stages.

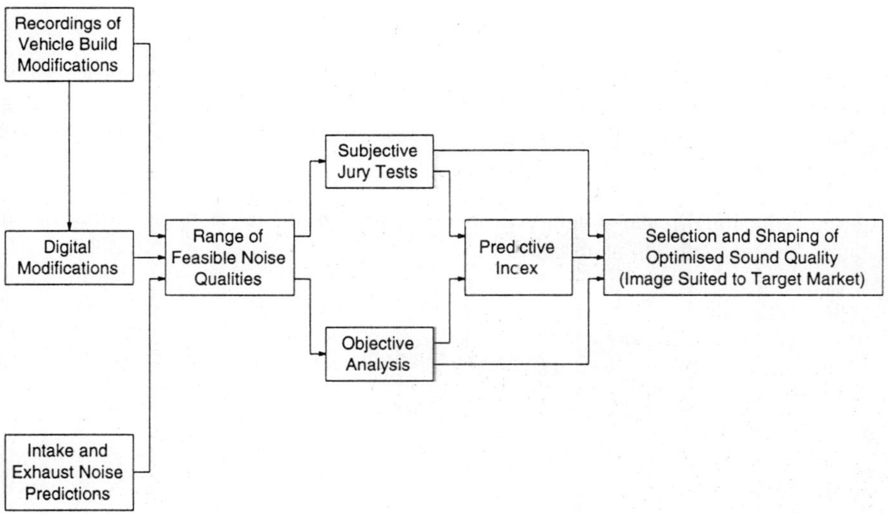

Figure 1 Overview of practical applications of sound quality analysis techniques

1. SUPPORTING IDENTIFICATION OF A ROAD-INDUCED NOISE PROBLEM

A medium sized passenger vehicle with a torsion beam rear suspension system was reported as having road-induced noise with an unpleasant character, which was both boomy and tonal in nature.

A subjective assessment method was used to determine which frequency bands contributed to the poor character. A set of filters was used to reduce the level of the noise by 12dB in 50 Hz wide bands with centre frequencies in the range 75 Hz to 425 Hz in 50 Hz steps. A panel of expert listeners was asked to decide which filters gave the most significant improvement to the unpleasant noise character.

After assessing the filtered noises, it was decided that the filters at 75 and 125 Hz gave significant subjective benefits to the boomy character, whereas those at 325 and 375 Hz significantly improved the tonal quality of the noise. Other filters had little or no perceptible effect.

Further developments of the vehicle were directed towards reducing resonance problems in these frequency regions, which were pinned down to specific components in the torsion beam using operating vibration and modal analyses.

2. SUPPORTING IDENTIFICATION OF ENGINE CRANK SHAFT RAP

A rapping noise from a V6 gasoline engine in a passenger car application was reported. The noise was thought to be crank shaft rap associated with excessive motion of the main bearings. It occurred under load at low engine speeds, and was more evident when the engine was cold.

A measure of knock intensity was developed to enable comparison between different engine builds during development. This was based on an energy integral of the sound pressure data for each individual engine cycle, bandpass filtered in the frequency range 500 to 1200 Hz in which the knocking noise was thought to occur. A subjective comparison was produced to validate the knock intensity measure. A bandpass filtered recording of the engine was compared with the same recording edited in the engine angle domain so that those engine cycles with the highest measured knock intensity were removed. Engine cycle noise removal was carried out using a cross-fading technique to avoid introducing any edit-derived clicks (see Figure 2).

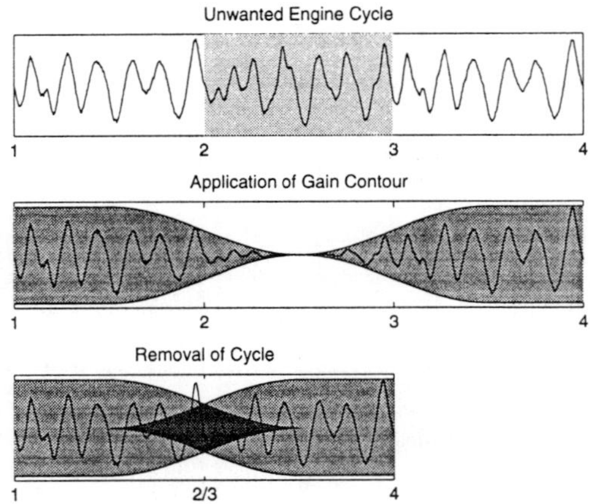

Figure 2 Removal of the noise of an engine cycle using cross-fading

The edited recording demonstrates a significant improvement in sound quality, confirming that those cycles with the highest measured knock intensity contributed significantly to the poor sound quality. This validates the measure, enabling its use with confidence during engine development.

3. PREDICTING INTAKE-INDUCED INTERIOR NOISE

A design for a new intake system, optimised for both performance and sound quality, was required for a small passenger vehicle.

Ricardo's computer-aided engineering code WAVE (1) was used to produce predictions of sound pressure data at the orifice of the modelled intake system before undertaking detailed design. Although a subjective assessment could have been carried out on the predicted orifice noise directly, it was considered important to hear the predictions as they would sound in the interior of the vehicle.

The acoustic transfer function between the orifice and the driver's ear position was measured. The inverse Fourier transform of this measurement was applied as a finite impulse response filter to the predicted intake orifice sound pressure data (see Figure 3). This calculation estimated the interior noise contribution from the modelled intake system.

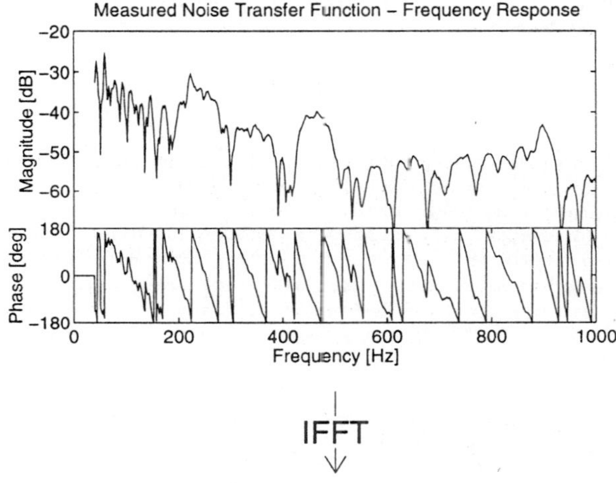

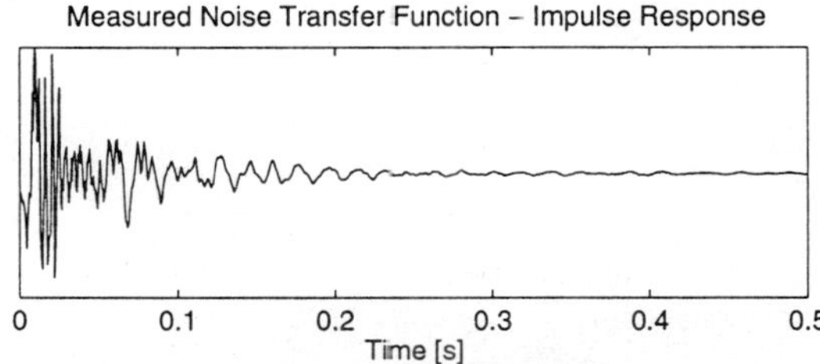

Figure 3 Measured noise transfer function

In addition, the interior noise of the vehicle with suppressed intake noise (using an infinite intake muffler) was recorded. This was assumed to have no contribution from intake orifice noise. Finally, the compound interior noise was calculated by summing the measured 'background' noise and the predicted and filtered intake noise. Full account of the engine cycle position was made to ensure that the phase relations between the different noise sources were correct.

This methodology gave an estimate of the noise of the modelled intake system in context inside the vehicle. Further reiteration of the modelling process would yield an intake system whose performance and interior noise character were optimised.

4. SETTING A NOISE CHARACTER TARGET

A V8 sports saloon vehicle manufacturer wanted to identify an exterior exhaust noise quality target that projected a sporty character. A range of feasible noise qualities was derived by preparing a series of vehicle builds with various exhaust system configurations. Recordings were made on a chassis dynamometer, with a binaural head placed one metre from the exhaust tail pipes. Some of these recordings were digitally modified to produce a wider range of potential noise characters.

The digital modifications were designed to alter the amount of 'rumble' character in the recordings, as this quality was considered important for the projection of a sporty image. Rumble is a form of amplitude modulation, and so is modified by changing the levels of the modulation sidebands, which occur at half and odd orders. Reducing the levels of the half and odd orders reduces the rumbly character, and vice versa. The level of a particular order is altered with a 'tracking' digital notch filter, in which the centre frequency of the notch is proportional to engine speed. Both reduced rumble and increased rumble recordings were generated. Waterfall plots of the unmodified and modified sounds are shown in Figure 4.

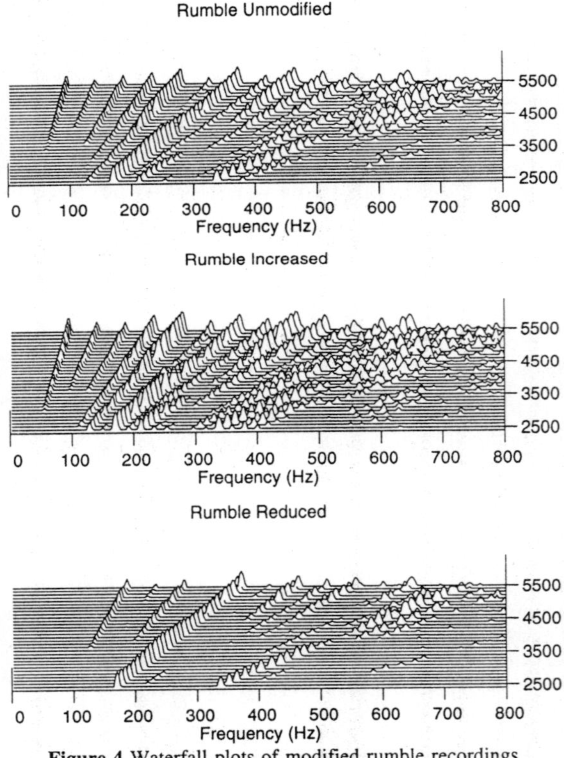

Figure 4 Waterfall plots of modified rumble recordings

The six noises finally selected for appraisal by the jury were arranged in a 6×6 'Latin square' format (2) as shown in Table 1.

Table 1. 6×6 Latin square format

A	B	F	C	E	D
B	C	A	D	F	E
C	D	B	E	A	F
D	E	C	F	B	A
E	F	D	A	C	B
F	A	E	B	D	C

This arrangement has two benefits. First, each noise is presented and therefore assessed several times, so that a mean rating can be obtained for each juror and the consistency of each juror assessed. Secondly, the test design is optimally 'balanced', that is each noise is preceded by each other noise at least once. Together, these two benefits overcome the influence of prior noises on current judgement. The jury was asked to rate the suitability of each noise for the intended application (in other situations another quality, such as annoyance, may be rated).

The noise selected by the jury as the most suitable to convey a sporty character had an even quality throughout the engine speed range, with some degree of rumble and very little flow-generated noise.

Having determined a noise character that was considered suitable for the intended vehicle application, development of the exhaust system was directed towards reproducing this character.

5. ASSESSING SPEECH INTELLIGIBILITY LEVELS

A truck fitted with a six-cylinder diesel engine and a power take-off facility produced an exterior noise quality that made speech communication difficult among vehicle users. A method was needed to assess the speech intelligibility levels around the truck.

A series of recordings was made using a binaural head located near the front wheel arch of the truck. A variety of modifications were applied to the truck, which was operated in several different conditions.

The vehicle noises were arranged in a jury test of the 'embedded speech' format (3), in which each noise was mixed with recordings of eight randomly selected, monosyllabic, phonetically balanced words. The number of words correctly recognised by each juror was counted as a measure of the speech intelligibility of each vehicle noise. A ranking of the vehicle noises according to the subjective speech intelligibility ratings gave useful information about the benefits of the various build and operating condition modifications.

Considerably higher levels of speech intelligibility were achieved with a lower engine speed. The power take-off facility caused a significant speech intelligibility problem. Reduction of airborne radiation from the powertrain gave significant improvements to speech intelligibility. A vehicle fitted with hydraulic electronic unit injectors gave poorer speech intelligibility levels than one fitted with mechanical unit injectors.

In addition, an objective index was derived to predict the subjectively generated speech intelligibility ratings. This index was a linear combination of a few objective metrics selected from the list in Table 2.

Table 2 A list of objective metrics

Overall Level	A, B, C, D, Linear SPL Zwicker Loudness
Speech Intelligibility	Speech Interference Level Articulation Index
Irregularity	Roughness Fluctuation Strength Kurtosis & Kurtosis Level Ricardo Impulsiveness Salience
Frequency Balance	Zwicker Sharpness SPL in bands
Tonality	Virtual & Spectral Pitch Harmonic Ratio

The index was constructed statistically using multi-variate regression (4) to determine the optimum linear coefficients, β_n, such that:

$$\text{Predicted Intelligibility} = \beta_0 + \beta_1 \times \text{metric}_1 + \ldots + \beta_n \times \text{metric}_n$$

This index is optimised by using a selection of metrics that provides the best fit to the subjective data, measured using correlation coefficient and mean squared error between the subjectively measured and the objectively predicted results. Further metrics are included in the index until the addition of another metric is statistically insignificant, thus determining the value of n (the 'dimensionality' of the index).

Where more than one metric is involved, the cross-correlations between all selected metrics are checked as a measure of their orthogonality. If the magnitude of all the cross-correlations are low, the metrics are likely to be orthogonal, removing redundancy from the index.

For the diesel truck, the optimum index for prediction of speech intelligibility levels was three-dimensional, as defined in Table 3.

Table 3 Definition of Optimum Index for Prediction of Speech Intelligibility Levels

suffix	β	metric
0	190	(constant)
1	1.91	an overall level metric
2	1.07×10^{-4}	an irregularity metric
3	0.375	another irregularity metric

This index is quick and easy to calculate in practise, so is ideally suited to the assessment of modifications to the vehicle during its development phase, replacing further time-consuming jury tests.

6. CONCLUSIONS

Sound quality analysis techniques are used increasingly in a wide range of practical applications. Setting a noise character target and assessing speech intelligibility are achieved using subjective jury testing. Objective prediction of these subjective test results is possible using statistically generated indexes. Modification of a sound's qualities using frequency, order, and angle domain editing helps to widen the possible range of noise characters in a subjective test, and supports identification of various noise problems.

Together, these techniques provide the tools necessary to give a clearly defined and systematic solution to a sound quality problem, and hence to convey a desired image to the consumer.

7. ACKNOWLEDGEMENTS

The author would like to thank the directors of Ricardo Consulting Engineers Ltd. for permission to publish this work.

8. REFERENCES

1. WREN, C.S., JOHNSON, O., Latest Simulation Techniques Applied to Intake and Exhaust System Design, ISATA, Aachen, 1994, 94ME044.
2. COCHRAN, W.G., COX, G.M., Experimental Designs, Wiley, 1957.
3. EGAN, J.P., Articulation Testing Methods, Laryngoscope, Vol. 58, 1948, pp 955 - 991.
4. MONTGOMERY, D.C., PECK, E.A., Introduction to Linear Regression Analysis, Wiley, 1982.

Experimental determination of low-frequency noise contributions of interior vehicle body panels in normal operation

J G van der LINDEN
LMS International, Belgium
P VARET
Renault-Creos, France

SYNOPSIS

Low frequency noise from engine-and wheel-vibrations often dominates the interior noise spectrum in vehicles. For the optimisation of vehicle bodies, there is an interest in quantification of the contribution of individual body panels to sound pressures at the passengers ear.

An experimental approach is presented which makes use of reciprocal acoustic transfer function measurements and surface acceleration measurements in normal road operation.

This method has been applied as a diagnostic tool to the interior noise of a four cylinder diesel engined van. On the basis of the results panels were tuned to reduce the noise.

NOTATION

P_r	=	sound pressure at ear position r (N/m^2)
P_{ri}	=	contribution of panel i at ear position r (N/m^2)
Q_i	=	panel i volume velocity (m^3/s)
H_{ir}	=	acoustic transfer from panel i to position r $(N.s/m^5)$
$\dot{X}_\perp$	=	local normal velocity on a panel (m/s)
o	=	direct experiment
$*$	=	reciprocal experiment

1. INTRODUCTION

Driver alertness, communication and comfort are all influenced by interior noise. Low noise therefore remains a major development issue. The noise inside vehicles has already been reduced significantly, but low frequency noise is persistent even in vehicles with well designed noise abatement.

Low frequency noise roots in the concept of the vehicle, engine suspension lay-out, aggregate structure, exhaust packaging and body structure. Within the given concepts, the system optimisation for low frequency noise is often a compromise between different vehicle properties, like handling-ride comfort-performance etc, costs are also involved.

Currently, the vehicle bodies may not be at the end of the optimisation. The interaction between the body and the cavity is a complex dynamic process. Hundreds of modes are present in the body structure below 400 Hz and the coupling between structural modes, trim modes and cavity modes make the system hard to describe in modal terms. A statistical approach on the other hand is still equally difficult, partially because of the low modal density in subsystems and partially because of the difficulties to estimate coupling loss factors.

Alternatively an experimental source-transfer system approach is presented, which can diagnose cavity-panel interaction. The experimental model is limited in its spatial definition but it allows optimisation as well as system studies for a better understanding .

The objective is a break down of interior noise into contributions of individual interior panels or parts of panels. The body and cavity are simplified to a source-transfer system. This approach, called Airborne Source Quantification (ASQ) , equally applies to other situations with predominant airborne noise transfer. The extracted quantities can also be drawn from models to allow hybrid modelling and subsystem optimisation.

2. METHOD OVERVIEW

The total interior boundary of the cavity is split up in panels or parts of panels. For each body panel the source strength is quantified in terms of volume velocity. The interior acoustic system, trim, isolation, absorption and cavity modes, are quantified by the acoustic transfer function between each panel and the sound pressure at the occupants ears.

The total interior noise can be described as a linear summation of contributions :

$$(1) \qquad\qquad P_r = \sum_i P_{ri} \qquad\qquad (N/m^2)$$

Each contribution is the product of a panel volume velocity "Q" and an acoustic transfer "H":

$$(2) \qquad\qquad P_{ri} = Q_i \cdot H_{ir} \qquad\qquad (N/m^2)$$

Airborne Source Quantification can thus quantify source strength, system sensitivity and noise contributions.

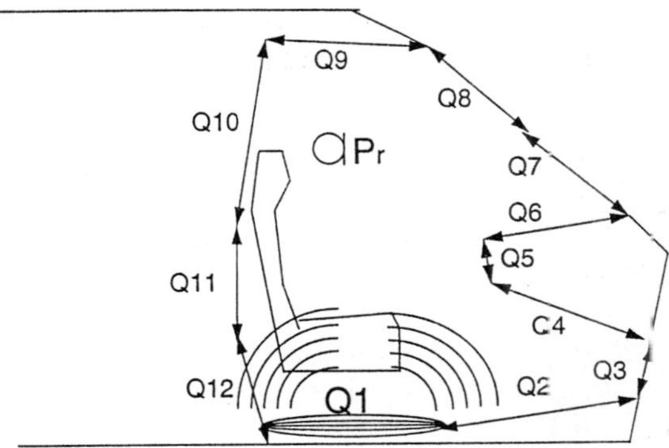

Fig. 1 : Vehicle interior, defined sources

3. ACOUSTIC TRANSFER

The acoustic transfer describes the pressure response at a certain position in the cavity due to the volume velocity of the defined panels. Normally it is necessary for the measurement of these acoustic transfer functions to build calibrated volume velocity sources into each panel. Only with a huge effort can such a measurement be achieved with acceptable accuracy. (Figure 1)

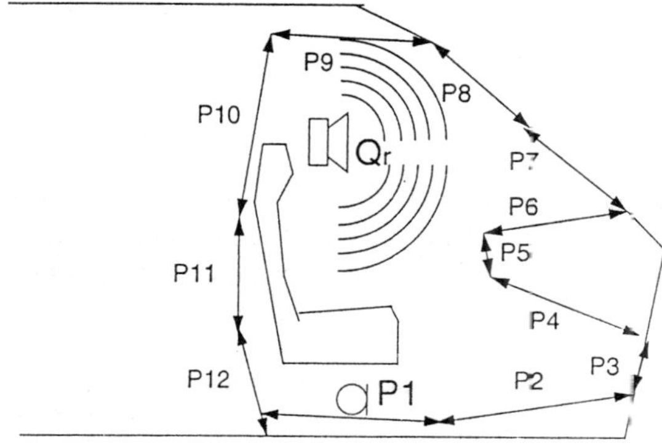

Fig. 2 : Vehicle Interior, reciprocal measurement

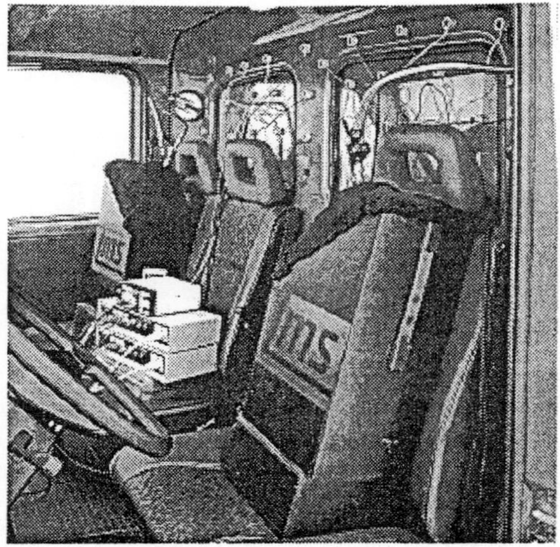

Fig. 3 : Instrumentation with low frequency volume velocity sources

We therefore make use of the reciprocity of the acoustic system to come to a feasible instrumentation. The reciprocity relation for the required body acoustic transfer functions:

$$(3) \qquad H_{ir} \quad = \quad \frac{P_r^0}{Q_i^0} \quad = \quad \frac{P_i^{\bullet}}{Q_r^{\bullet}} \qquad (Ns/m^5)$$

It is far more practical to place a sound source at the ear position and measure sound pressure at the different panels than to insert a sound source at each panel. For reciprocal measurements a sufficiently small omnidirectional and well calibrated volume acceleration sound source is placed at the passenger ear position and small microphones are placed at all panels.(Figure 2) In this way the measurement effort is also reduced as there are usually essentially more panel positions than ear positions.

4. VOLUME VELOCITY

As a result of the acoustic transfer function method it has becomes necessary to quantify the operational source strength of the panels also as the volume velocity of a point source on the centre of each panel. This seems to deviate from the actual vibrating panels. However, the sound field of a point source with volume velocity Q on a surface will equal the sound field of a piston with surface normal velocity V, and area ds, as long as the largest dimension of the area ds remains much smaller than the acoustical wavelength.

For panel i the sound pressure at an arbitrary position in the sound field can be expressed as :

$$(4) \qquad P_{ri} = \int_{S_i} H_{ir}\, \dot{x}_{\perp}\, ds \qquad (N/m^2)$$

But, when the acoustic transfer "H" is near to constant over the area S_i, then it can be placed outside the integral and the equivalent volume acceleration becomes :

$$(5) \qquad\qquad P_{ri} = H_{ir} \int_{S_i} \dot{X}_\perp \, ds \qquad\qquad (N/m^2)$$

The panels surface normal velocity can be measured with light accelerometers, optically or by other means, whichever is most appropriate to the situation.

The surface velocity sampling density is of course critical. This will depend on the expected bending mode shape in the target frequency range and also on the surface geometry.

For the contribution calculation it is irrelevant whether the surface normal velocity data are measured in vehicle operation or during artificial excitations. The surface normal velocity can also relate to structural modes derived from measurements or from numerical models. In all situations a contribution ranking is possible.

5. CASE STUDY, DIESEL VAN

The object of investigation is a large volume van with a four cylinder turbo diesel engine. The interior noise spectrum is balanced in most driving conditions except for two rpm ranges. Under full load condition the second order component of the engine noise becomes dominant around 1400 rpm throughout the cabin and at around 2600 rpm at the drivers side. (Figure 4). The second order component of the forces from the engine under full load, which act onto the body structure, have already been optimised. But for the two critical rpm ranges the noise transfer functions at the force input locations have maxima at the corresponding frequencies of 45 and 90 Hz.(Figure 5) It was targeted to identify the panel contributions to the interior sound pressure over the entire rpm range and subsequently to devise modifications to reduce the body sensitivities.

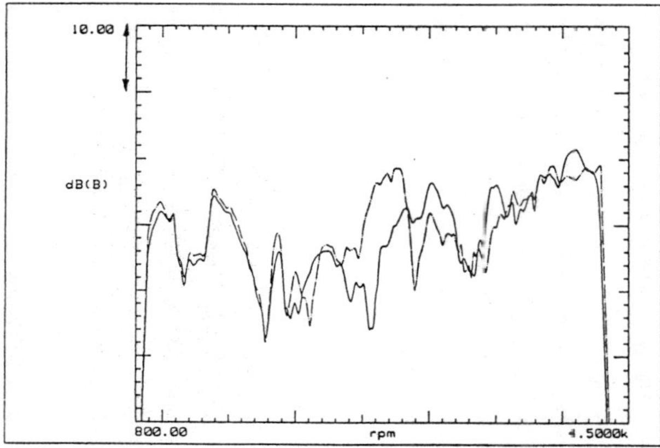

Fig. 4. Second order component of the sound pressures at the driver(solid line) and of the passenger position(dash-dot line), all in third gear and full load.

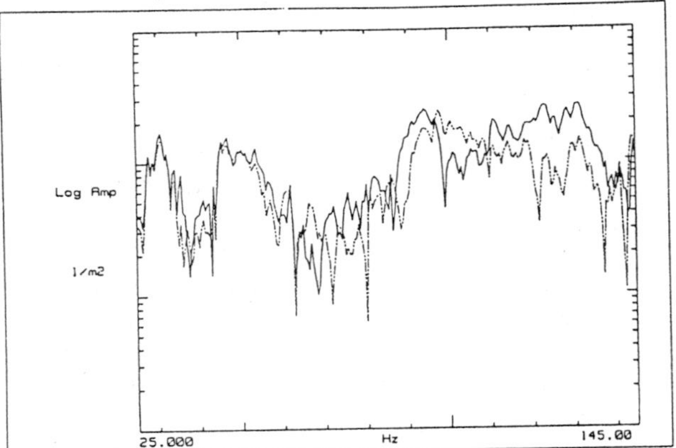

Fig. 5: Vibro acoustic FRFs from engine suspension to ear positions. Driver position in solid and passenger position in dash-dot line.

6. MEASUREMENT IN OPERATION

The cabin surrounding panels were split up in sub areas and instrumented with accelerometers. Measurements were performed on a smooth long road, in batches, until all 2nd order acceleration curves, phase referenced, were available. The samples spatial density was aiming at a ± 150 Hz upper limit. The measured acceleration data also allow a limited running mode animation of the panel normal motion. It is, however, most often impossible to identify the dominant panels on the basis of the panel motion only. The differences in acoustic transfer for different panels are to large to disregard. On top of that it is common that the operational deformations show many phase changes between panel patches, which make interpretation difficult. But panel motion animation does give some understanding of the type of panel modes, which are active in the frequency range of interest.

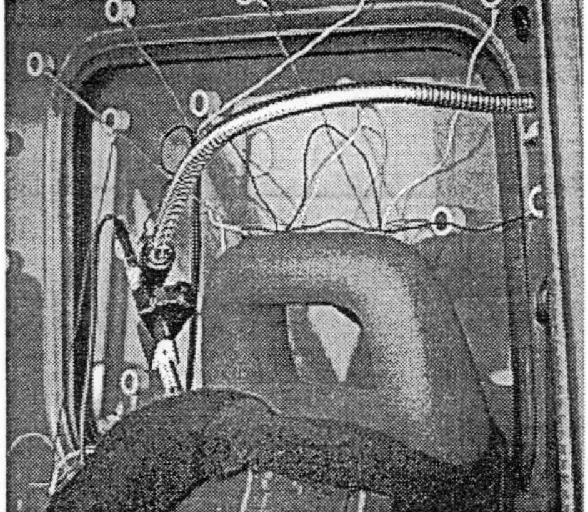

Fig. 6 : Panel normal velocity, instrumentation with accelerometers

7. ACOUSTIC TRANSFER FUNCTIONS

Two calibrated volume velocity sources were placed at the ear positions of the driver and the passenger. And again in batches all sound pressures on the panel sub areas were measured. Uncorrelated burst random excitation was used to drive the sources. Figure 7 shows two examples of measured pressure over volume velocity transfer functions. These transfer functions are to be used directly for the ASQ panel contribution analysis.

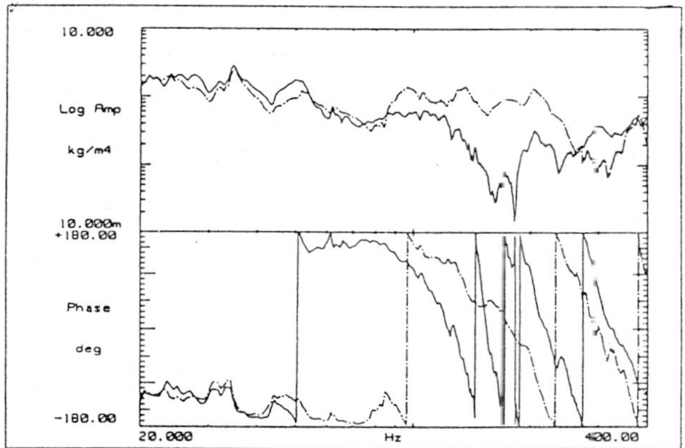

Fig. 7 : Acoustic FRF examples from upper (dash-dot) and lower(solid) rear wall to the drivers ear position

An additional acoustic modal analysis was made on the basis of the same acoustic excitation. For this purpose additional pressure measurements in the cabin volume were added to the transfer function database. An acoustic mode is shown in figure 8.

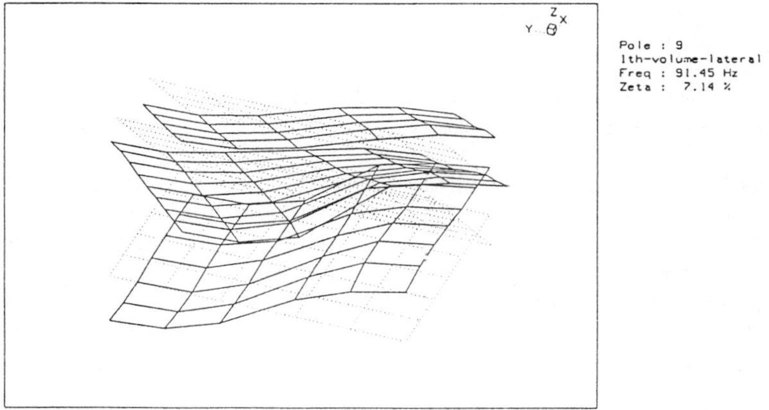

Fig. 8 : First lateral acoustic mode, seen from rear right top of the cabin.   Pressure is animated as vertical displacement

8. PANEL CONTRIBUTION ANALYSIS

The panel accelerations were converted to source volume velocity by a discretisation of equation 5.

$$(6) \qquad Q_i = \sum_i \dot{X}_i S_i \qquad\qquad (m^3/s)$$

All panel contributions were calculated, summed and the sum of all panel contributions is compared to two direct measured 2nd order sound pressure curves. The comparison of the identified noise and the measured noise shows that the analysis remains within the spread of the measurements up to 125 Hz. Both amplitude and phase are well approximated.(Figure 8) The direct total interior sound pressures were taken from the first and from the last road measurement of this series.

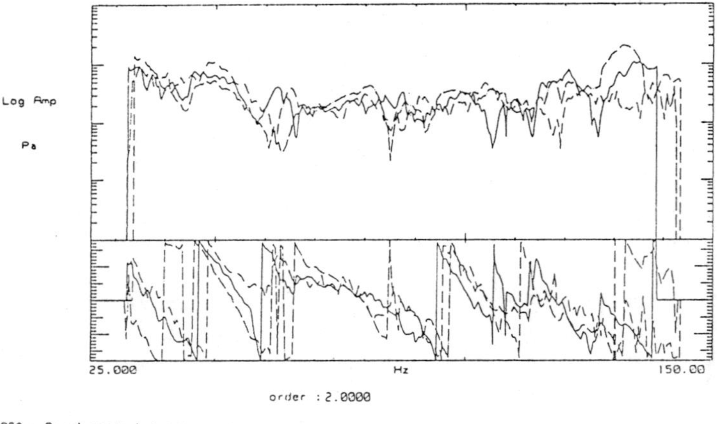

Fig. 9 : Measured second order noise on the road (solid and long-dash lines) and identified sum of contributions (dash-dot line)

The individual panel contributions are the essential part of the analysis. Figures 9 and 10 show examples of identified contributions under third gear full throttle acceleration.

It was most interesting to find out that, also for this airborne process, the phase of each contributions is equally important as the amplitude of a contribution. It was found that absorbent action of the panels is easily recognised from the panel analysis by the investigation of the phase. One of the panels showed very high normal acceleration values as well as high volume velocity values. But its contribution to the interior noise is out of phase with the interior noise and effectively this panel is acting as an absorber in most of the rpm range. Only at the panels first bending mode at 50 Hz is actively contributing to the interior noise.(Figure 10)

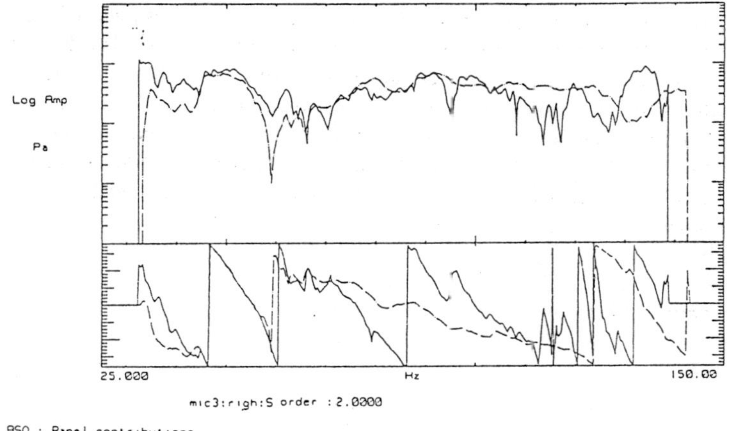

Fig. 10 : Measured second order noise (solid) and a major contributing and absorbing panel (dashed).

9. PANEL MODIFICATION

Two rpm ranges were targeted for improvement.

Around 1500 rpm, well below the first acoustic modes, the wind shield was the dominant source. It exhibits a global bending combined with some deformation at the boundaries. A modification lowered the bending mode in frequency to 34 Hz. This lowered the second order maximum at 1500 rpm in the entire vehicle cabin, but it has increased the 2nd order level around 1100 rpm. The 1500 rpm range is considered more important because of more frequent use in steady state driving conditions, whereas 1100 rpm full load is untypical.(Figure 11)

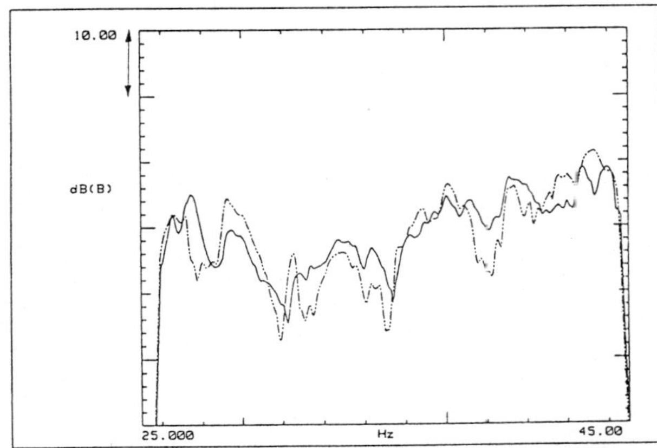

Fig. 11 : Effect of the panel tuning on the second order noise. Original dashed and modified solid.

Around 2400 rpm, the first lateral acoustic mode determines the acoustic response of the cabin. Left and right footweld floor areas were the dominant contributors. These already stiff floor sections can not be modified easily. A considerable modification of the basic vehicle structure would be required to reduce the volume velocities of the floor areas. Therefore an other type of solution was tried. The first plate bending modes of two smaller panels were tuned to 90 Hz. The small panels were well placed near the pressure maxima of the acoustic mode and could therefore act efficiently as acoustic vibration dampers. This reduced the maximum in the 2nd order curve as can be seen in fig. 12

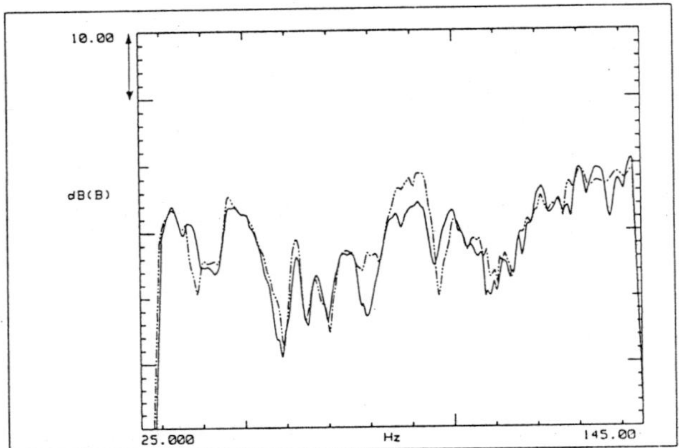

Fig. 12 : Effect of tuned panel absorbers on the second order noise, original dashed and modified in solid

Both modifications provided only an optimisation of the second order curves. But the solutions are suitable for rapid implementation and the comfort is improved by smoothing the 2nd order noise curves.

10. CONCLUSIONS

A methodology is presented, called Airborne Source Quantification (ASQ), to analyse contributions of panels to the interior noise, producing both phase and amplitude information.

The application on a diesel van shows the possibility to identify dominant contributions as well as absorbent action of panels. A set of modifications was developed on the basis of the diagnosis results, which lowered the interior noise at critical rpm ranges.
Airborne Source Quantification (ASQ) is modular in several respects;
- source and transfer are separated
- experimental and modelling data can be combined.

This methodology is undergoing further development and is also applied to other airborne noise areas.

The laser torsional vibrometer – optimum use in the presence of lateral vibrations

T J MILES, M LUCAS, and S J ROTHBERG
Loughborough University of Technology, UK

SYNOPSIS

Torsional vibration of rotating shafts can contribute significantly to vehicle noise and vibration levels. The Laser Torsional Vibrometer has been developed for accurate non-contact measurement of the torsional oscillation of rotating components, with advantages over conventional techniques. Its optical geometry offers inherent immunity to lateral shaft vibration in most cases but the instrument is sensitive to specific types of lateral shaft vibration under certain circumstances. The importance of this cross-sensitivity is demonstrated with experimental results from engine applications, allowing the specification of optimum configurations to ensure effective insensitivity to all lateral vibration.

NOMENCLATURE

d Beam separation

f_{beat} Photodetector output beat frequency

$\hat{i}$ Unit vector defining the direction of the incident laser beams

k_g Geometric constant

$\overline{k}_g$ Mean value of k_g

Δk_g Amplitude of fluctuations in k_g

N Rotation frequency

ΔN Amplitude of fluctuations in rotation frequency

$\hat{r}$ Unit vector perpendicular to $\hat{i}$ and $\hat{z}$

t Time

$\hat{z}$ Unit vector defining direction of shaft rotation

α Incidence angle in $\left(\hat{i},\hat{z}\right)$ plane

β Incidence angle in $\left(\hat{r},\hat{z}\right)$ plane

$\Delta\alpha$ Amplitude of angular motion in $\left(\hat{i},\hat{z}\right)$ plane (yaw)

$\Delta\beta$ Amplitude of angular motion in $\left(\hat{r},\hat{z}\right)$ plane (pitch)

λ Wavelength of laser light

μ Refractive index

ω Angular frequency of angular lateral motion

1. INTRODUCTION

Torsional vibration is an important measurement in vehicle NVH studies. The interest in its effects has increased as design trends concentrate on energy efficiency and power to weight ratio, driven by legislation and customer demands. Numerous problems in automotive NVH applications include torsional oscillation of engine crankshafts, excitation of transmission system driveshafts, idling rattle and gearbox backlash noise. The adverse effects of torsional resonance are increased mechanical shear stresses, component wear and fatigue, together with reduced passenger compartment comfort. Fundamental work has identified relationships between crankshaft torsional vibration and noise for diesel engines (1) and the application of torsional vibration absorbers and dampers to automotive engines has been studied with the aim of achieving NVH improvements (2, 3). Despite the obvious importance, measurement of torsional vibration has not been widely used in machinery diagnostics as traditional measurement systems require engine modifications or have performance limitations.

The Laser Torsional Vibrometer (LTV) (4) achieves accurate measurement of the torsional oscillation of rotating components, offering the advantage over conventional techniques of non-contact use. Operation utilises the Doppler frequency shift in laser light back-scattered from a moving target. The LTV performs successfully on shafts of arbitrary cross-section and the beams may be incident on either the side or face of the shaft. Its unique optical geometry offers immunity to axial shaft motion and radial shaft vibration such as the cylindrical whirl orbit of figure 1a and provides substantial advantages over previous non-contact techniques (5, 6). However, it has previously been demonstrated (7) that the instrument is sensitive to lateral vibration in a whirl orbit where the target shaft rotation vector undergoes a change of direction, such as the conical whirl orbit shown in figure 1b.

The effect of this lateral vibration on the LTV measurements has been confirmed experimentally and this paper compares this effect with measured torsional vibration levels. Measurement of the lateral vibration of a resiliently-mounted engine confirmed that these errors can become significant under specified operating arrangements. Comparison of simultaneous torsional vibration measurements from two LTVs set at different incidence angles verified the occurrence of this effect. Guidelines are therefore proposed for accurate assessment of genuine torsional vibration by optimisation of.the optical geometry used for reciprocating engines in which the crankshaft undergoes angular lateral vibration.

2. EFFECTS OF LATERAL SHAFT VIBRATION

2.1 Theory of Operation

Figure 2 shows the LTV beams incident on the side of a shaft of arbitrary cross-section, rotating at frequency N about an axis defined, in the absence of angular lateral vibration, by the unit vector $\hat{z}$. Two equal intensity parallel beams of separation d impinge on the shaft surface in a direction defined by the unit vector $\hat{i}$ at points A and B. Recombination of the incident beams on the photodetector produces a beat frequency equal to their difference frequency, with torsional vibration seen as a fluctuation in this beat frequency. Defining $\hat{r}$ as a unit vector perpendicular to both $\hat{z}$ and $\hat{i}$, then α is their included angle, as shown in figure 3a. Inspection of figure 3b allows definition of β as the angle between the plane of the laser beams and unit vector $\hat{r}$. The general expression for the beat frequency f_{beat} in terms of a geometrical constant k_g is then (7, 8):

$$f_{beat} = k_g N \qquad \{1a\}$$

where:
$$k_g = \left(\frac{4\mu\pi d}{\lambda}\right)\cos\beta \sin\alpha \qquad \{1b\}$$

It can be seen from equation {1b} that the main factors over which the operator has control in determining f_{beat} are the laser beam separation, d, and the set-up angles α and β. Practical limits to the usable range of angles α and β are determined by the collection of sufficient scattered light from the target and the minimum value of f_{beat} which can be demodulated (4). During hand-held use α and β can be affected (8) and tripod mounting, while not essential, is preferable. It is also important to note that angular lateral vibration of the shaft affects the angles α and β causing them to become time dependent. Shaft angular lateral vibration is considered as the combination of rotations in the $(\hat{i},\hat{z})$ plane, affecting α, and the $(\hat{r},\hat{z})$ plane, affecting β, classified as yaw and pitch respectively (9).

2.2 Target Shaft Lateral Vibration

Previous work by the authors (7) has developed appropriate theory and experimental validation to highlight the significance of the instrument's sensitivity to lateral vibration in a whirl orbit where the shaft rotation vector undergoes a change of direction and consequently α and β are time dependent. Examples include the conical whirl orbit shown in figure 1b, vibrations of flexible shafts and rocking of an engine under test on its mountings. This paper further demonstrates this sensitivity with experimental results from engine applications.

Optimum operation of the instrument is achieved when the plane of the incident laser beams is perpendicular to the rotation axis, then $\alpha = 90°$ and $\beta = 0°$ to give:

$$k_g = \left(\frac{4\mu\pi d}{\lambda}\right) \qquad \{2\}$$

When aligned in this way sensitivity to variations in N is maximised and small variations in α and β will have little effect on instrument output. Use of the instrument at other incidence angles increases the sensitivity to angular motion errors, in addition to reducing the overall sensitivity, but is often demanded by restricted access.

When lateral vibration causes changes in k_g the effect on the LTV output will be indistinguishable from that resulting from a genuine torsional vibration. The beat frequency resulting when a steadily rotating shaft undergoes angular lateral vibration can be considered equivalent to that resulting when the shaft rotates with a mean speed N and a torsional vibration ΔN. From equations {1a & b} variations in k_g and N can be written:

$$\bar{k}_g(N + \Delta N) = \left(\bar{k}_g + \Delta k_g\right)N \qquad \{3a\}$$

where $\bar{k}_g$ and Δk_g are the mean value and the amplitude of fluctuations in k_g respectively. This is more conveniently expressed in the form of an apparent speed fluctuation ratio:

$$\frac{\Delta N}{N} = \frac{\Delta k_g}{\bar{k}_g} \qquad \{3b\}$$

2.3 Derivation of Errors Induced by Angular Lateral Vibrations

Equation {1b} can be written in terms of variations in α and β with time, designated $\Delta\alpha(t)$

and $\Delta\beta(t)$ respectively for the yaw and pitch motions, and expressed in the form of equation {3b}, neglecting powers higher than first order:

$$\frac{\Delta N}{N} = \frac{\Delta\alpha(t)}{\tan\alpha} - \Delta\beta(t)\tan\beta \qquad \{4\}$$

From the above equation the magnitude of any error component will depend principally on the incidence angles α and β, and the angular motions $\Delta\alpha(t)$ and $\Delta\beta(t)$. In order to quantify the effect of each motion, the problem is simplified by considering them separately and obtaining appropriate forms of equation {4}. If $\Delta\alpha(t) = \Delta\alpha\sin\omega t$, where $\Delta\alpha$ is the amplitude of the angular motion and $\Delta\beta = 0$, equation {4} can be written:

$$\frac{\Delta N}{N} = \left(\frac{\Delta\alpha}{\tan\alpha}\right)\sin\omega t \qquad \{5\}$$

Full evaluation reveals the presence of components at integer multiples of ω but the first order component dominates. The same approach allows consideration of the condition when β varies with time and $\Delta\alpha = 0$ but operation of the LTV with $\beta \neq 0°$ is uncommon and the errors induced by this angular lateral vibration will not be considered further here.

2.4 Experimental Validation of Theory

This theory was verified using a test-rig to experimentally simulate these angular vibrations (7), offering accurate angular tilt motion of a rotating shaft section within the required frequency range. The amplitude and frequency of the motion were readily controllable, together with the rotational speed, providing a large dynamic range for experimentation.

Figure 4 shows typical experimental and theoretical results for the yaw or $(\hat{i},\hat{z})$ plane lateral motion error as a function of α with $\beta = 0$ and $\Delta\beta = 0$. Testing at further rotation speeds confirmed the validation of equation {5}. One set of values was obtained with the laser beams incident on the side of the shaft and a second set were taken from viewing the end face of the shaft, a technique often used with the LTV in engine measurements where a pulley or gearwheel often provides the only convenient access. The instrument noise floor, which shows harmonic speckle noise peaks (10), was also monitored and an average $\Delta N/N$ value of 3×10^{-4} found. The optimum operating point is confirmed as $\alpha = 90°$.

It is now necessary to consider the magnitude of these errors due to the angular lateral vibration found in practice, relative to the instrument noise floor and genuine levels of torsional vibration.

3. ENGINE VIBRATION MEASUREMENTS

3.1 Engine Torsional Vibration

Measurement of torsional vibration was made from the crankshaft pulley of a four-cylinder 2.0 litre direct injection diesel engine under full load. Expressed as $\Delta N/N$, for direct comparison with the values obtained in the previous section the second order exhibited the most severe torsional vibration, with a maximum $\Delta N/N$ of 0.021 (602m° displacement) at

1300rpm. Smaller torsional resonance effects were clearly apparent for the eighth and sixth orders. Measurements were also made from the petrol engine discussed below and for the speed range covered under load a second order maximum $\Delta N/N$ of 0.049 (1420m°) at 1040rpm was recorded. A minimum level of significant torsional vibration at any order for each engine was estimated to be $\Delta N/N = 0.001$.

3.2 Engine Yaw and Pitch

Actual angular lateral vibrations from an automotive application, $\Delta\alpha(t)$ and $\Delta\beta(t)$, were assessed to determine limits within which the operation of the LTV will give accurate measurement. Tests on a resiliently mounted four-cylinder 2.0 litre fuel injection petrol engine measured the angular motion of the engine block and gave a peak yaw amplitude of 36m° for second order vibration when running under load and full throttle at 545 rpm. The maximum pitch amplitude was approximately 10m° second order at 775rpm.

3.3 Comparison of LTV Measurements

To highlight discrepancies in LTV measurements due to angular lateral vibration of the target shaft, two LTVs were set up to measure simultaneously the torsional vibration of the crankshaft pulley of the engine used in section 3.2. One was set at $\alpha = 20°$ (very sensitive to angular lateral vibration) and the other at $\alpha = 60°$ (marginally sensitive to angular lateral vibration), arranged to be incident in the yaw plane as defined previously. Calibration was carried out across the speed range for both instruments and the measured second order torsional vibration results are shown in figure 5.

Above 1400rpm the values for both instruments are approximately equal. However, as the speed decreases the difference between the measurements increases, as a result of angular lateral vibration of the rotating target. Since there is no method to find the phase between the torsional and lateral motions it is not possible to give an accurate figure for genuine angular lateral vibration. However, the difference between the measurements at speeds around 1000rpm is estimated to correspond to a genuine angular lateral vibration of the crankshaft pulley of the order of 100-200m°. This is approximately four times the value measured for motion of the engine block but is of appropriate magnitude when compared to published data on consideration of the effects of crankshaft bending vibrations (11).

4. DISCUSSION OF RESULTS

For accurate assessment of torsional vibration it is necessary to identify operating regimes where the sensitivity of the LTV to vibrations $\Delta\alpha(t)$ and $\Delta\beta(t)$ is insignificant. The experimental data discussed in section 4 is presented in figure 6 to give a clear indication of the significance of errors expected. The typical instrument noise floor level specifies a minimum above which an angular lateral vibration error becomes significant. The maximum and minimum levels of $\Delta N/N$ measured during the engine tests are included. Comparison is made with the error induced by angular lateral vibrations as predicted by equation {7}, from the maximum yaw previously determined, $\Delta\alpha = 36$m°, and a value of crankshaft pulley angular motion, $\Delta\alpha = 150$m°, estimated from the simultaneous LTV measurements.

This upper value of yaw angle amplitude is used to define the operating limits for use of the LTV on reciprocating engines in which the target shaft experiences these levels of angular lateral vibration. In figure 6 it can be seen that, for α values down to approximately 80°, any signal component induced by angular lateral vibration is of similar or smaller magnitude than the instrument noise floor. From this incidence angle down to 70° the error components are of measurable magnitude and this range should be used with caution. With $\alpha < 70°$ the magnitude of signal components induced by angular lateral vibration becomes significant and it is recommended that use of the LTV in this range is avoided if possible. The target shaft motion should be analysed if accurate LTV measurements are to be obtained in this range.

Further work is considering optical configurations of the instrument to remove this sensitivity to angular lateral vibration and to achieve measurement of the actual crankshaft bending vibrations.

5. CONCLUSIONS

Operation of the Laser Torsional Vibrometer has been discussed, with theoretical and experimental examination of the influence of lateral shaft motion on its measurements, advancing the understanding of previous work.

Measurements from diesel and petrol engines demonstrated the torsional vibration which can arise in practice. The angular lateral vibration of a resiliently mounted four-cylinder petrol engine was assessed, through measurement of the engine block motion, and typical maximum yaw levels were found to be 36m°. Subsequent comparison of two simultaneous measurements of torsional vibration from this engine, taken at different LTV incidence angles, highlighted discrepancies between the instruments. Yaw vibration of the order of 100-200m° was estimated from this data. This increase in angular lateral vibration compared to previous values was attributed to the angular motion of the crankshaft pulley.

A comparison was made between the relative magnitudes of angular lateral vibration induced components, genuine torsional vibration and the LTV noise floor and particular operating regimes were highlighted where the sensitivity to angular lateral vibration becomes of concern in this typical application. These guidelines are important if use in these configurations is necessitated by shaft access limitations. Error components are potentially very significant when $\alpha < 70°$. In the range $70° < \alpha < 80°$ the error components are of measurable magnitude and care is required in the interpretation of data. For $80° < \alpha \leq 90°$ error components were comparable to the instrument noise floor level and operation is effectively immune to all lateral vibration, allowing reliable and accurate measurement of torsional vibration.

REFERENCES

(1)   OCHIAI K & NAKANO M, Relationship Between Crankshaft Torsional Vibration and Engine Noise. <u>Transactions of the Society of Automotive Engineers</u>, 1979, Vol. 88, Section 2, Paper 790365, pp1291-1298.

(2)   BAKER JM, BAZELEY G, HARDING R, HUTTON DN & NEEDHAM P, Refinement Benefits of Engine Ancillary Dampers. Proceedings of the Institution of Mechanical Engineers, Vehicle NVH and Refinement, May 1994. Paper C487/041/94, pp149-158.

(3)   BOROWSKI VJ, DENMAN HH, CRONIN DL, SHAW SW, HANISKO JP, BROOKS

LT, MIKULEC DA, CRUM WB & ANDERSON MP, Reducing Vibration of Reciprocating Engines with Crankshaft Pendulum Vibration Absorbers. <u>Transactions of the Society of Automotive Engineers</u>, 1991, Vol. 100, Section 2, Paper 911876, pp376-382.

(4)   HALLIWELL NA, PICKERING CJD & EASTWOOD PG, The Laser Torsional Vibrometer: A New Instrument. <u>Journal of Sound and Vibration</u>, 1984, Vol. 93, No. 4, pp588-592.

(5)   HALLIWELL NA, PULLEN HL & BAKER J, Diesel Engine Health: Laser Diagnostics. <u>Transactions of the Society of Automotive Engineers</u>, 1983, Paper 831324, pp3.986-3.944.

(6)   VANCE JM & FRENCH RS, Measurement of Torsional Vibration in Rotating Machinery. <u>Transactions of the ASME Journal of Mechanisms, Transmission and Automation in Design</u>, 1986, Vol. 108, pp565-577.

(7)   MILES TJ, LUCAS M & ROTHBERG SJ, The Laser Torsional Vibrometer: Successful Operation During Lateral Vibrations. 15th ASME Biennial Conference on Mechanical Vibration and Noise, Boston, Massachusetts, 17-21 September 1995. (To be presented.)

(8)   HALLIWELL NA & EASTWOOD PG, The Laser Torsional Vibrometer. <u>Journal of Sound and Vibration</u>, 1985, Vol. 101, No. 3, pp446-449.

(9)   COLLACOTT RA, <u>Vibration Monitoring and Diagnosis</u>. 1979, (George Goodwin, London) Chapter 2, pp16-18.

(10)   ROTHBERG SJ, BAKER JR & HALLIWELL NA, Laser Vibrometry: Pseudo-Vibrations. <u>Journal of Sound and Vibration</u>, 1989, Vol. 135, No. 3, pp516-522.

(11)   ALBRIGHT MF & STAFFELD DF, Noise and Vibration Refinement of the Ford 3.8 Liter Powertrain. Proceedings of the 1991 Noise and Vibration Conference, Society of Automotive Engineers P-244, Paper 911073, pp291-305.

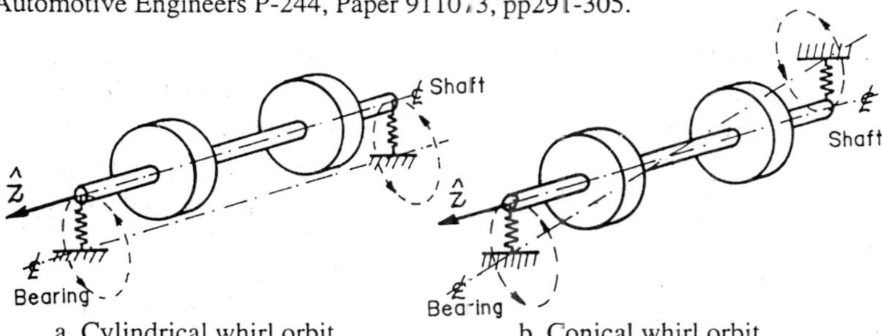

a. Cylindrical whirl orbit   b. Conical whirl orbit

Fig 1 Rigid-rotor lateral vibration modes for a symmetrical rotor

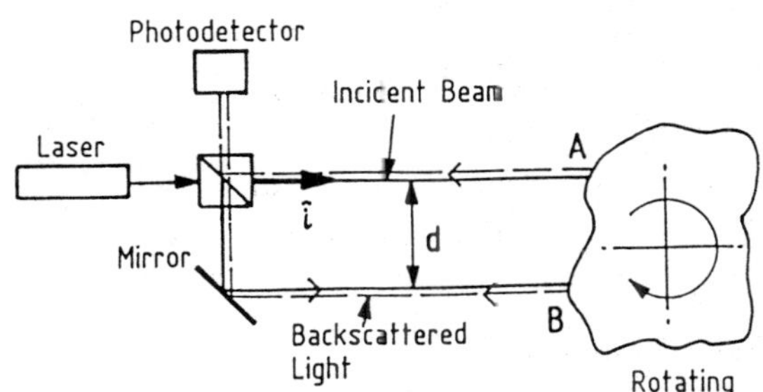

Fig 2 Optical geometry of the Laser Torsional Vibrometer

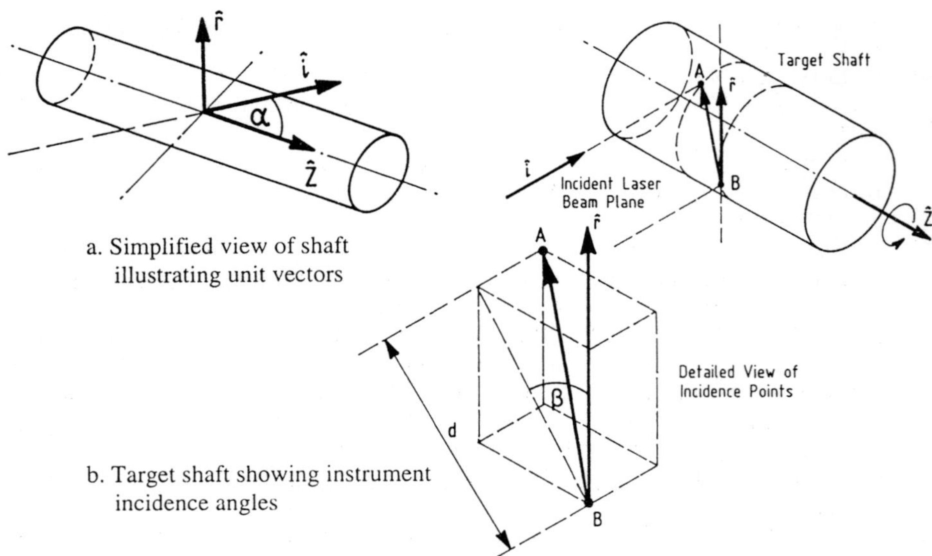

a. Simplified view of shaft
illustrating unit vectors

b. Target shaft showing instrument
incidence angles

Fig 3 Torsional Vibrometer instrument-target configuration

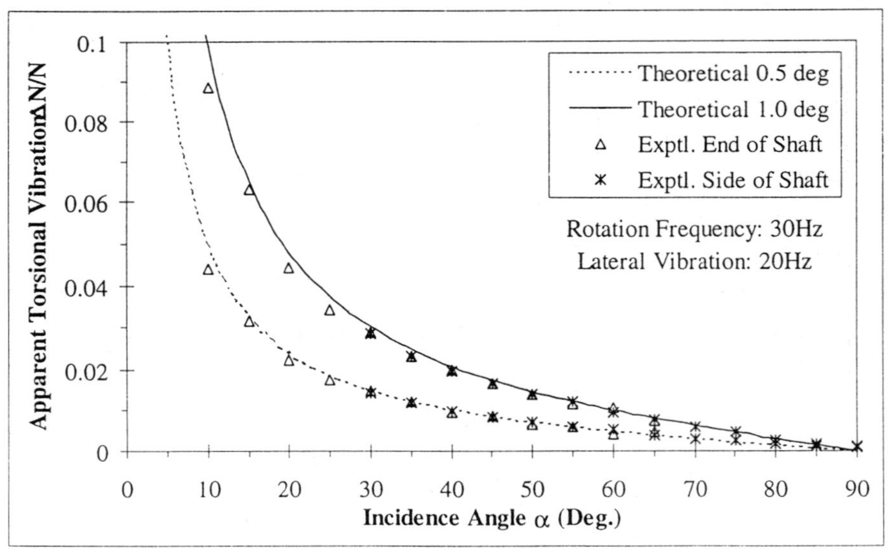

Fig 4 Yaw or $\left(\hat{i}, \hat{z}\right)$ plane lateral motion error

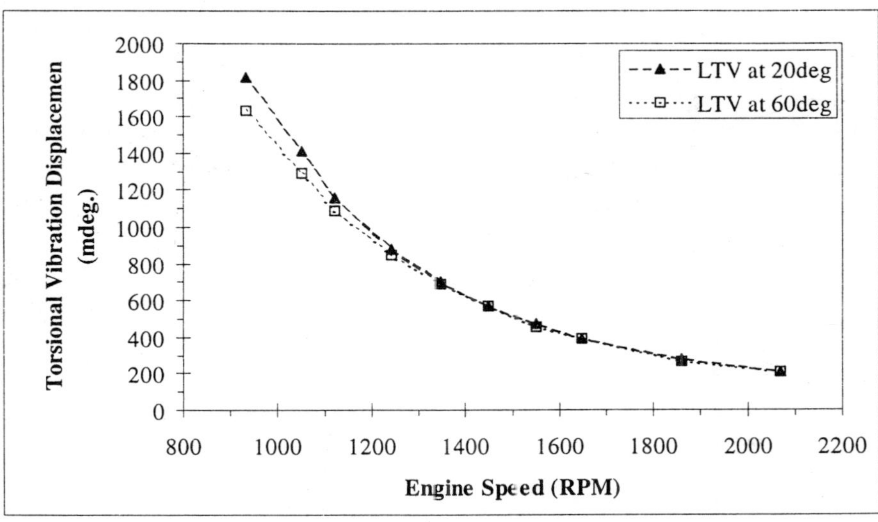

Fig 5 Comparison of simultaneous LTV measurements at different incidence angles

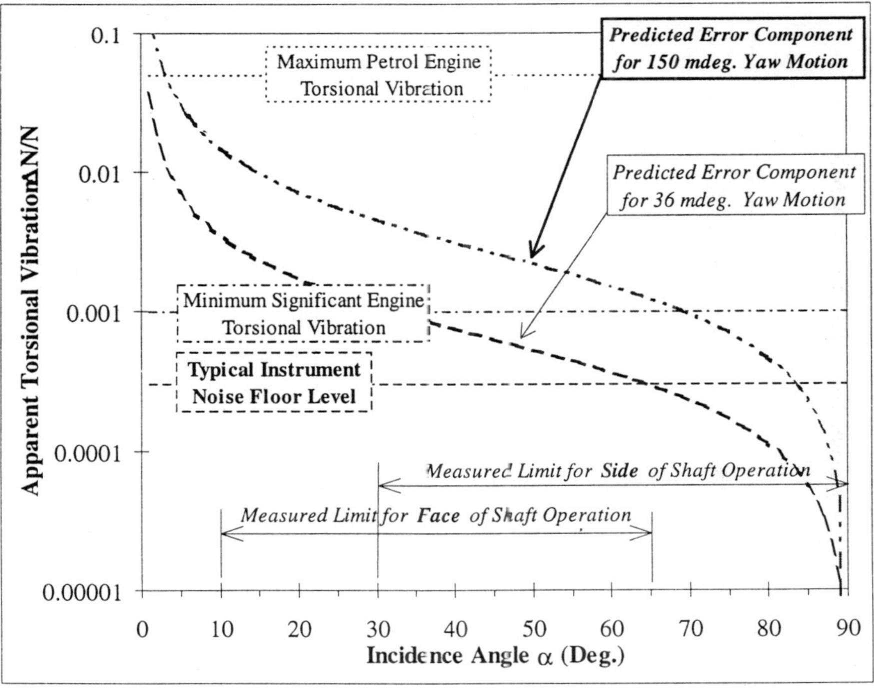

Fig 6 Guidelines for successful operation of Laser Torsional Vibrometer with minimisation of lateral vibration errors

Suspension and Ride Handling

Full vehicle modelling and the requirements for accurate handling simulations

M V BLUNDELL MSc
Coventry University, UK

SYNOPSIS

The influence of suspension modelling on the accuracy of simulation output has been investigated by comparing the results from several computer models with data obtained during instrumented vehicle testing. The complexity of the models has been varied to include a vehicle where the suspensions are represented by sliding masses, a vehicle where the suspensions are modelled as swing arms, a vehicle model based on roll stiffness and a final detailed model containing the geometry of the true vehicle suspension linkages. The results were compared with the test data to establish the influence of suspension model complexity on handling simulation accuracy.

1 INTRODUCTION

The vehicle handling simulations described in this paper have been performed using the ADAMS (*Automatic Dynamic Analysis of Mechanical Systems*) computer program. Multi body systems analysis programs such as ADAMS (1) have become well established within the automotive industry where they are used mainly to study the behaviour of suspension systems or to simulate the ride and handling performance of a new design using (2-6). Full vehicle models in ADAMS may contain elements representing all the suspension linkages and the flexible bushes making up the connections. Obtaining this data can be difficult and time consuming and could lead to a situation where the simulation model can not respond to the pace of design activity.

The objective of this study was to investigate the influence of simplified suspension models on the accuracy of calculated outputs from vehicle handling simulations. This work has resulted due to experiences in industry where highly detailed models containing in excess of one hundred degrees of freedom have been used for handling simulations.

In terms of developing the sort of full vehicle models described in this paper it is worth noting the comments provided in reference (7):-

"Models do not possess intrinsic value. They are for solving problems. They should be thought of in relation to the problem or range of problems which they are intended to solve. The ideal model is that with minimum complexity which is capable of solving the problems of concern with an acceptable risk of the solution being "wrong". This acceptable risk is not quantifiable and it must remain a matter of judgment. However, it is clear that diminishing returns are obtained for model elaboration."

The concept of refining a model for a particular analysis is well established in finite element modelling and can be considered as a two stage process. The first stage is to define an idealisation for the model. This involves making experienced judgments such as how to constrain a model, apply loads, exploit symmetry or select element types. The result is an idealisation or in other words a model which is 'ideal'. The second phase is more straightforward and involves deciding on the size and distribution of elements throughout the model. This is referred to as the discretisation. Typically an analyst would refine the distribution of elements until the calculated stresses converged on a realistic value. Many finite element programs can now automate this process. For the multibody systems analyst involved in setting up a vehicle model for a handling simulation the process is not so straightforward. There is no discretisation as such. All decisions are in fact in the area of setting up an idealisation. The modelling issues will be fundamental and may include how to represent the suspension, roll bars, whether to include body flexibility, to model bushes as linear, nonlinear or not at all. The selection of a tyre model is another major issue.

This paper attempts to address this issue by considering a range of suspension modelling strategies and comparing outputs from a typical vehicle handling simulation. The effects of varying a tyre model have not been considered at this stage but is a planned investigation at a future date.

For this work the tyre force and moment characteristics were represented using the default ADAMS Fiala tyre model which has the advantage of requiring only ten input parameters, but is recognised as being unsuitable for combined slip situations such as braking in a turn. For the full vehicle model used to represent the actual linkages the complexity of the model could also be varied according to the manner used to represent the compliant bush connections. This was investigated separately using suspension (quarter) models and is described in the next section.

2 SUSPENSION MODELLING STUDY

For the vehicle considered in this paper the front and rear suspensions were modelled as separate units and then analysed moving through the full range of vertical movement between the bump and the rebound positions. The output from this type of simulation is mainly geometric and allows results such as camber angle or roll centre position to be plotted graphically against vertical wheel movement. An investigation was carried out to establish the influence of modelling the bushes on the calculated suspension outputs most likely to influence vehicle handling behaviour.

For each suspension system three types of model have been considered:

(a) Modelling bushes as nonlinear
(b) Modelling bushes as linear
(c) Modelling with rigid joints (kinematic analysis)

A secondary objective from this phase of work was to establish for both front and rear suspensions the positions of the instant centre and the roll centre. The positions of the instant centre were used later as pivot points for a simple full vehicle handling model where the suspensions were represented by single swing arms. The roll centres were used for a simple full vehicle handling model based on roll stiffness. Both of these simplified full vehicle handling models are described later in this paper. The modelling of the front suspension only is described here. The modelling of the suspension system using bushes is shown in Figure 1. The rigid joint model is shown in Figure 2. The upper link is attached to the body using a connection which is rigid enough to be modelled as a revolute joint. Bushes were used to model the connection of the lower arm and the tie bar to the vehicle body. Bushes were also used to model the connections at the top and bottom of the damper unit. Where the tie bar is bolted to the lower arm a fix joint has been used to rigidly connect the two parts together. This joint removes all six relative degrees of freedom between the two parts creating in effect a single lower wishbone. The modelling issue raised here is that rotation of the lower arm takes place about an axis through the two bushes but that the bushes are not aligned with this axis. As rotation takes place the bushes must distort in order to accommodate the rotation. The modelling of these connections as nonlinear, linear or as a rigid joint was therefore investigated to establish the effects on suspension geometry changes during vertical movement.

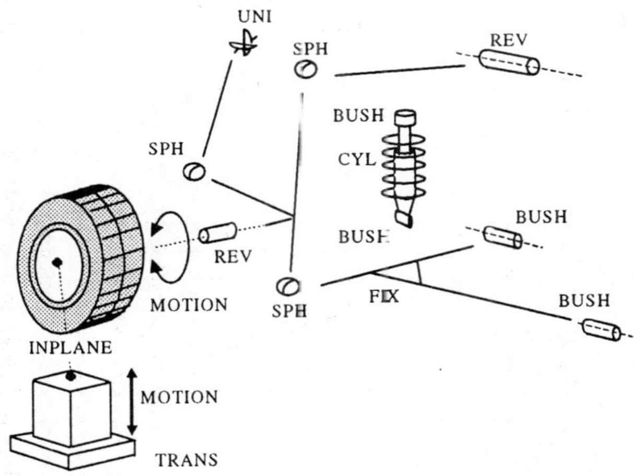

Figure 1 Modelling the front suspension using bushes

In order to produce a zero degree of freedom model for this suspension the bushes at the top and bottom of the strut have been replaced by a universal and a spherical joint. The bushes connecting the lower arm and the tie rod to the vehicle body have been replaced by a single revolute joint with an axis aligned between the two bushes as shown in Figure 2.

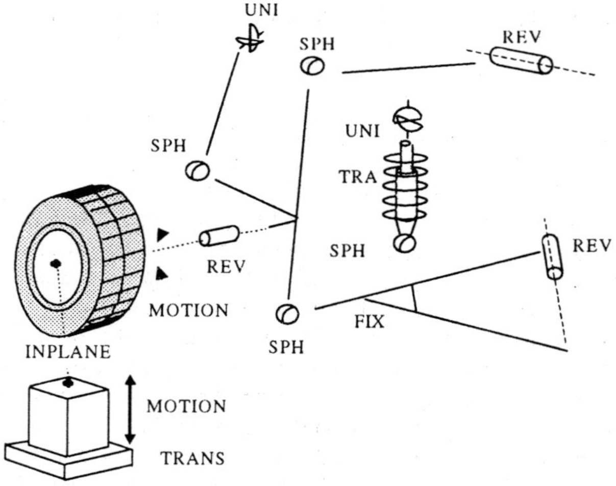

Figure 2 Modelling the front suspension using rigid joints

The analysis involves moving the suspension through the full range of vertical movement using a jack part which imparts motion to the wheel centres. The analysis results are plotted graphically with the vertical displacement (Bump movement) plotted on the X-axis. For the front suspension it was possible to compare the ADAMS results with measured data provided by the manufacturer. The results shown here are for the camber angle variation in Figure 3 and the steer change in Figure 4. The following line types have been used to identify the curves of suspension outputs for each model:

Rigid joints   · · · · · · · · · · ·   Nonlinear bushes   ————————
Linear bushes   — — — — — —   Test data   — · — · — · — · —

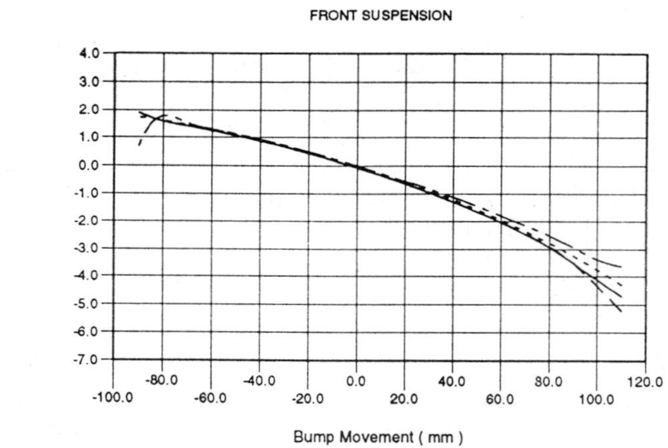

Figure 3 Camber angle with bump movement

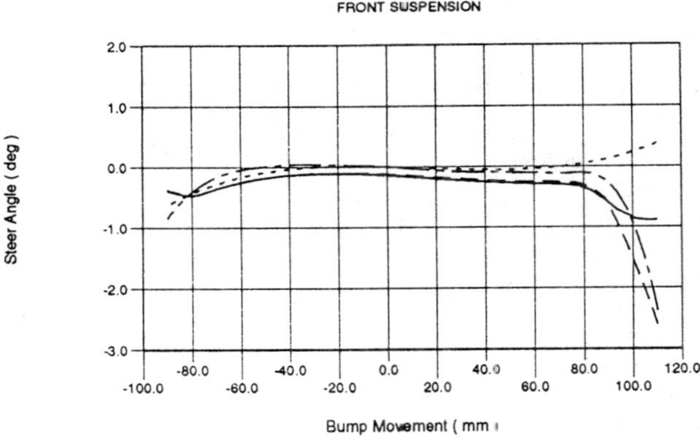

Figure 4 Steer angle with bump movement

Comparison of the results for both the front and rear suspension models shows that there is good agreement between models using rigid joints, linear bushes and nonlinear bushes, with the exception of steer change in the rear suspension where a rigid jointed model was unable to accurately represent this. It is noticeable with the front suspension that the plots begin to deviate when approaching the full bump or full rebound positions. This is due to the forces building up in the bump stop or rebound stop which are reacted through the suspension to the bushes which results in the changes in suspension geometry as shown in the plots. This effect is not present in the models using rigid joints which have zero degrees of freedom and where geometry changes are entirely dependent on the position and orientation of the joints.

The use of the nonlinear model will significantly increase the effort required to model the vehicle. The modelling of a linear bush requires only a reasonable amount of extra data input in ADAMS when compared with the rigid joint model provided that bush data is available to the analyst.At this stage it would appear that for a handling model of this vehicle based on modelling linkages either the linear bush or nonlinear bush models could be incorporated but that the rigid joint model would be unsuitable due to the lack of accuracy in representing the rear suspension bump-steer characteristics. The use of the nonlinear model will significantly increase the effort required to model the vehicle. This is evident from Table 1 which compares the raw ADAMS data inputs required to model the connection of the front suspension lower arm to the vehicle body. As can be seen there is a significant amount of data required to model a nonlinear bush whereas the modelling of a linear bush requires only a reasonable amount of extra data input when compared with the rigid joint model.

From Table 1 it is clear that the modelling of nonlinear bushes has a significant impact on the preparation of the input deck particularly as there are often in excess of twenty bushes in any one full vehicle model. There is also a great deal of extra effort which should be documented. For the nonlinear bushes in these models the characteristics were entered in the form of X-Y pairs making up a nonlinear spline. The values used here are based on test data provided by a vehicle manufacturer. It is important to check that the spline which ADAMS fits through the data is consistent with the test figures. For each of the nonlinear splines the data should be plotted and checked. In some cases the spline fit is poor leading to an oscillatory characteristic in the spline. In these cases it is necessary to fit additional points in the test data to ensure a smooth curve fit.

Table 1 ADAMS data input for a joint, linear bush and nonlinear bush

NONLINEAR BUSH	LINEAR BUSH	JOINT
BUSH/16,I=1216,J=0116 ,K=0,0,0 ,KT=0,0,500 ,C=35,35,480 ,CT=61000,61000,40 GFORCE/16,I=1216,JFLOAT=011600,RM=1216 ,FX=CUBSPL(DX(1216,0116,1216),0,161)\ ,FY=CUBSPL(DY(1216,0116,1216),0,161)\ ,FZ=CUBSPL(DZ(1216,0116,1216),0,162)\ ,TX=CUBSPL(AX(1216,0116),0,163)\ ,TY=CUBSPL(AY(1216,0116),0,163)\ ,TZ=0.0\ SPLINE/161 ,X=-1.8,-1.5,-1.4,-1.22,-1.123,-1.0,-0.75,-0.5,-0.25,0,0.25,0.5,0.75,1.0,1.123,1.22,1.4,1.5,1.8 ,Y=15350,10850,9840,6716,5910,5059,3761,2507,1253,0,-1253,-2507,-3761,-5059,-5910, ,-6716,-9840,-10850,-15350 SPLINE/162, ,X=-5,-4,-3,-2.91,-2.75,-2.5,-2,-1.5,-1,-0.5,0,0.5,1,1.5,2,2.5,2.75,2.91,3,4,5 ,Y=7925,3925,1925,1790,1626,1450,1136,830,552,276,0,-276,-552,-830 ,-1136,-1450,-1626,-1790,-1925,-3925,-7925 SPLINE/163, ,X=-0.22682,-0.20939,-0.19196,-0.17453,-0.1571,-0.13963,-0.10472,-0.06981 ,-.03491,0,0.03491,0.06981,0.10472,0.13963,0.1571,0.17453,0.19196,0.20939,0.22682 ,Y=241940,198364,160018,125158,93387,75415,52951,35702,18453,0,-18453,-35702 ,-52951,-75415,-93387,-125158,-160018,-198364,-241940	BUSH/16,I=1216,J=0116 ,K=7825,7825,944 ,KT=2.5E6,2.5E6,500 ,C=35,35,480 , CT=61000,61000,40	JO/16,REV,I=1216,J=0116

It is also very easy to make an error when entering such large amounts of nonlinear data in the ADAMS input deck. The plotting of nonlinear data is therefore a necessary activity in terms of the quality assurance of the model but very time consuming. On this basis the full vehicle model using linkages to model the suspension is represented using the linear bush model.

3 VEHICLE MODELLING

The three models discussed here are shown schematically in Figure 5. In the first of the simplified models the suspensions were modelled as lumped masses, connected to the vehicle body by translational joints which allow vertical sliding motion. Spring and damper forces act between the suspensions and the body. The front wheel knuckles were modelled as separate parts connected to the lumped mass suspension parts by revolute joints. The steering motion required for each manoeuvre was achieved by applying time dependent rotational motion inputs about these joints. Each road wheel was modelled as a part connected to the suspension by a revolute joint. The tyre forces and moments were calculated automatically using the Fiala tyre model. The roll bars were modelled using two parts connected to the vehicle body by revolute joints and connected to each other by a torsional spring located on the centre line of the vehicle.

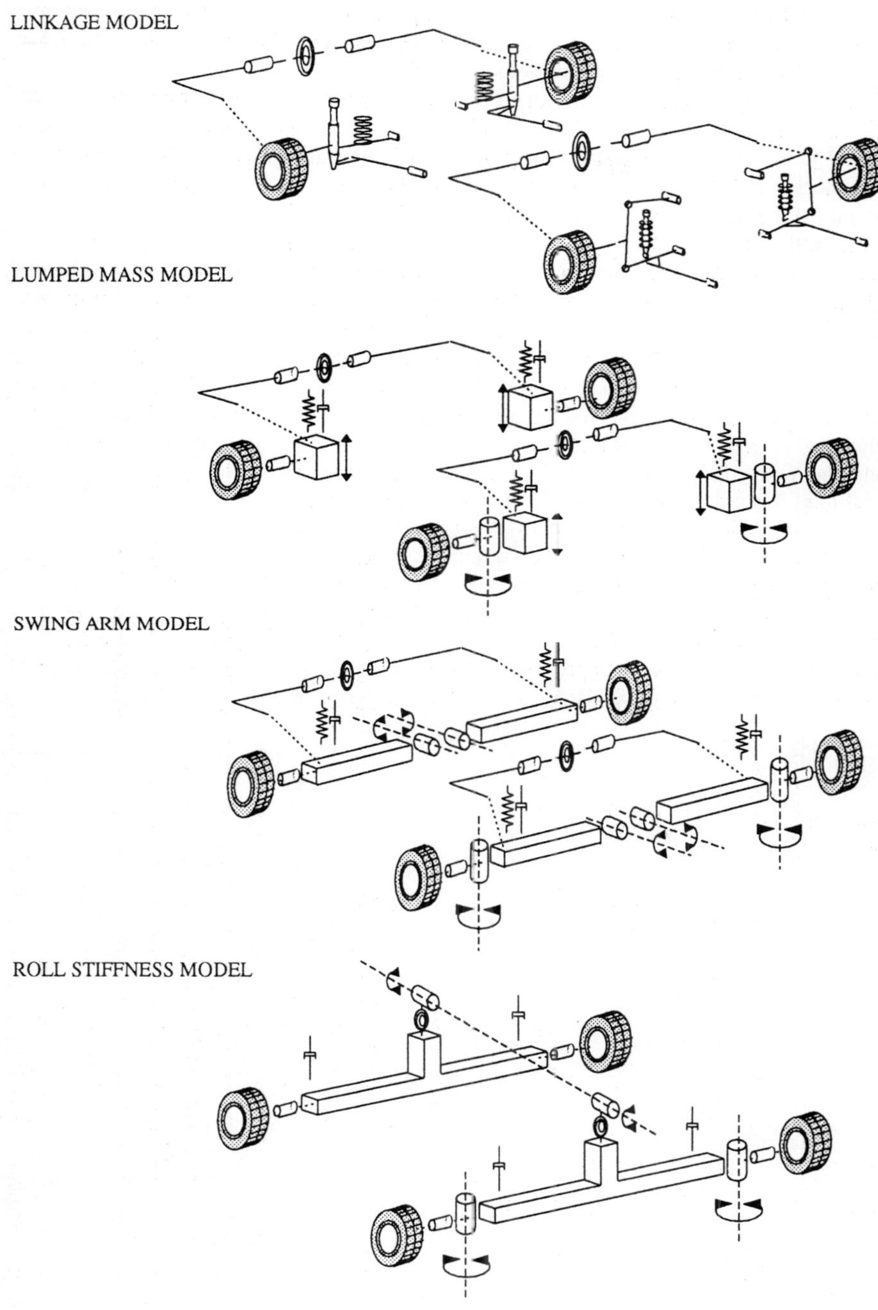

Figure 5 Modelling of suspension systems

Each roll bar part was connected to the suspension using an inplane joint primitive which allowed the vertical motion of the suspension to be transferred to the roll bars and hence produce a relative twisting motion between the two sides.

The swing arm model was developed from the lumped mass model by using revolute joints to allow the suspensions to 'swing' relative to the vehicle body rather than using translational joints which only allow sliding motion to take place. The revolute joints were located at the instant centres of the actual suspension linkage assembly. These positions were found by modelling the suspensions separately as described in Section 2.

The roll stiffness model was a further simplification treating the front and rear suspensions as rigid axles connected to the body by revolute joints at the roll centres. The roll centre positions were obtained from the study described in Section 2. A torsional spring was located at the front and rear roll centres to represent the roll stiffness of the vehicle. As there was no measured data for roll stiffness it was necessary to determine the roll stiffness of the front and rear suspension elements separately. The procedure used to find the roll stiffness for the front suspension elements involved the development of a model as shown in Figure 6. This model included the vehicle body which was constrained to rotate about an axis aligned through the front and rear roll centres. The vehicle body was attached to the ground part by a cylindrical joint located at the front roll centre and aligned with the rear roll centre. The rear roll centre was attached to the ground by a spherical joint in order to prevent the vehicle sliding along the roll axis. A motion input was applied at the cylindrical joint to rotate the body through a given angle. By requesting the resulting torque acting about the axis of the joint it was possible to calculate the roll stiffness associated with the front end of the vehicle. The front suspensions were modelled using the suspension model where bushing characteristics were treated as linear. The springs were also included as was the complete front roll bar model. The road wheel parts were not included nor were the tyre properties.

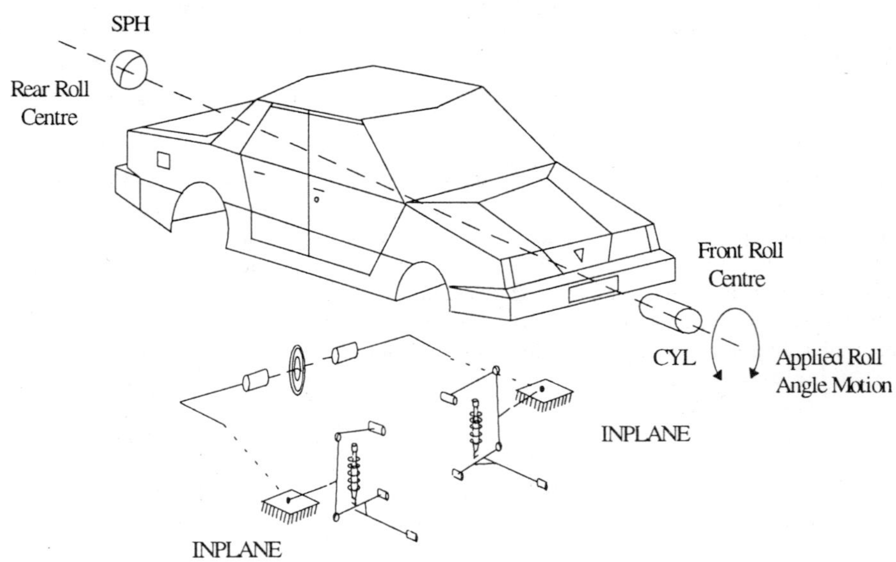

Figure 6 Determination of front end roll stiffness

The wheel centres on either side were constrained to remain in a horizontal plane using inplane joint primitives. Although the damper force elements were retained in the suspension models they have no contribution as the roll stiffness was determined using quasi-static analysis. The steering system, although not shown in Figure 6, was also included in the model. A motion input was used to lock the steering in the straight ahead position during the roll simulation. Hand calculations were performed to check the approach and showed good agreement. The rear roll stiffness was obtained in the same manner.

4 DISCUSSION

For each of the models described here simulations were carried out to represent sudden fixed steering inputs. These manoeuvres are also referred to as control response simulations and are based on the measured handwheel angles obtained during vehicle testing. In each case the steering input is suddenly applied, held constant for a period of about five seconds and then returned to the straight ahead position. The handwheel angles used during testing were available in the form of a time history plot. One way of representing this in ADAMS could be to set up the curve as a set of xy pairs and allow ADAMS to use a cubic spline fit to recreate the steering inputs during the simulation. Examination of the measured data, however indicated that the steering inputs were smooth enough to use a mathematical 'STEP' function to simply ramp the steering input on and off over the measured time intervals. In all the simulations the vehicle speed was set at 100 kph. Each analysis in ADAMS was performed over a ten second period with one hundred output steps. A typical view of the animated graphics is shown in Figure 7. The results are plotted in Figure 8 to Figure 15 and show the variation in lateral acceleration and body roll angle for a 60 degree steering input. The ADAMS results are plotted with a solid line and compared with the test data plotted with a dashed line.

The test data shows fluctuations which may be due to slight oscillations in the steering input during the test which are not represented in the simulation. Other causes may be vibration due to imperfections in the test track or aerodynamic forces which are also not included in the analysis. Additional test data has been obtained for a lane change manoeuvre at a 100 kph where the problem with oscillations is not so evident. The next phase of work will compare the simplified models with this test data with the hope that this might further identify any weaknesses or strengths with the simplified models Examination of the results reveals little variation between the models with the exception that the lumped mass model appears to roll about one degree further than the other three models. The suspension models described here appear to work well for this particular vehicle and this simulation. Due to the wide range of vehicle types, function and suspension arrangements it is probably not possible to define a definite modelling strategy for vehicle handling simulation that is transferable from vehicle to vehicle. It is also worth noting that the simplified models which use data such as instant centres and roll centre positions are based on the static suspension geometry and migrate during vehicle roll. This being the case caution is required when using models for extreme manoeuvres such as limit cornering. Having considered the simplified models described here a further step would be to include another modelling approach which is gaining favour and is described in (8). This approach makes use of suspension derivatives to model the change in position and attitude of the wheel relative to the vehicle body. Simple models such as those described here are likely to be most useful at an early stage of vehicle design. During the later stages in design vehicle models could be extended to include all the linkages. These models would be useful for durability analysis looking at the diffusion of loads from the tyre to road interface through the suspension and into the vehicle body.

The effects of model complexity on computer simulation time were recorded and are shown in Table 2. The times are for simulations running on a Viglen 4DX266 personal computer. As can be seen the effect of model size is not significant. The simulation time does not increase at the same ratio of model size. Simulation times are more normally effected by a single force equation which is highly nonlinear and requires very small integration step sizes.

Table 2 The effect of model size on computer simulation time

Model	Degrees of freedom	Number of Equations	CPU Time (s)
Linkage	78	961	146
Lumped Mass	14	429	108
Swing Arm	14	429	93
Roll Centre	12	265	68

Acknowledgments

The authors would like to acknowledge the support of Rover Group and SP Tyres (UK) Ltd for the work which is described in this paper.

References

(1) RYAN R. ADAMS - Multibody Systems Analysis Software, Multibody Systems Handbook/Werner Schielen (Editor), (1990)
(2) ORLANDEA, N. and CHASE, M. Simulation of a vehicle suspension with the ADAMS computer program, SAE paper 770053 (1977)
(3) RAI, N. S. and SOLOMAN, A. R., Computer simulation of suspension abuse tests using ADAMS, SAE paper 820079 (1982)
(4) ANTOUN, R. J., HACKERT, P. B., O'LEARY, M. C., SITCHEN, A. Vehicle dynamic handling computer simulation - Model development, correlation, and application using ADAMS, SAE paper 860574 (1986)
(5) BLUNDELL M.V. Full vehicle modelling and simulation using the ADAMS software system, Autotech '91, IMechE, (1991)
(6) BLUNDELL M.V., PHILLIPS B.D.A. and MACKIE A. The modelling and simulation of Suspension Systems, Tyre Forces and Full Vehicle Handling Performance, ICSE '94, Coventry (1994)
(7) SHARP, R.S. Computer codes for road vehicle dynamics models, *Autotech '91, IMechE*, Birmingham, November, (1991)
(8) SCAPATICCI, D., COELI, P., MINEN, D. ADAMS implementation of synthetic trajectories of the wheels with respect to the car body for handling manoeuvres simulations. Proceedings of the 8th European ADAMS News Conference, Munich, October, (1992)

Figure 7 Vehicle handling simulation graphic output

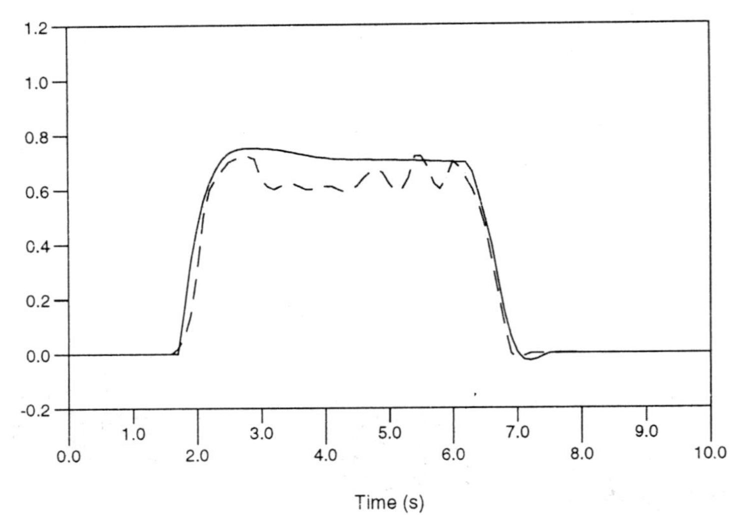

Figure 8 Lateral Acceleration

Figure 9 Roll Angle

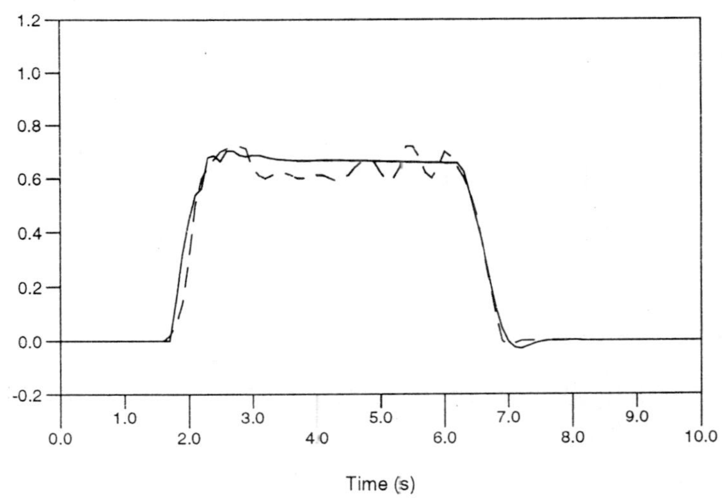

Figure 10 Lateral Acceleration

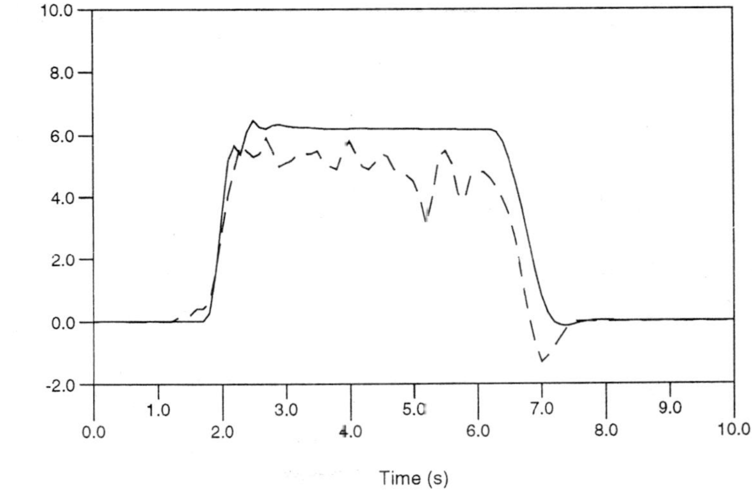

Figure 11 Roll Angle

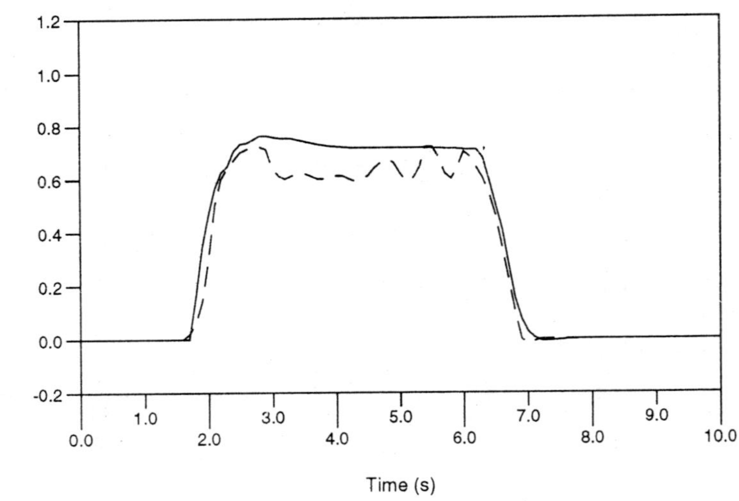

Figure 12 Lateral Acceleration

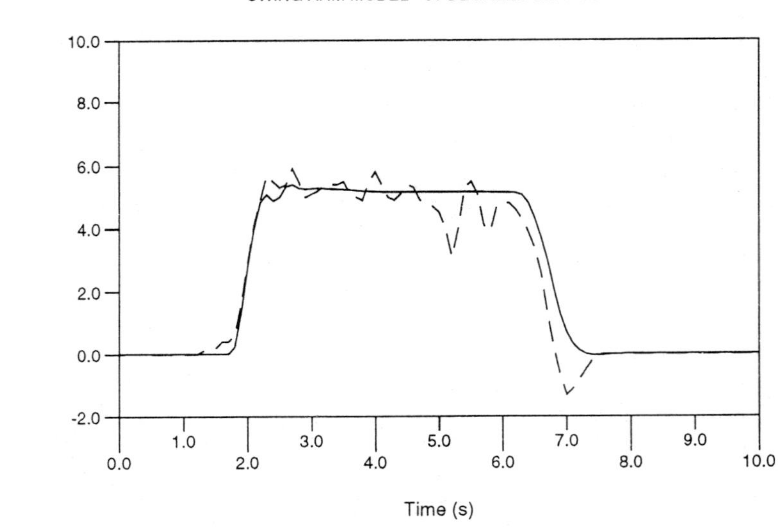

Figure 13 Roll Angle

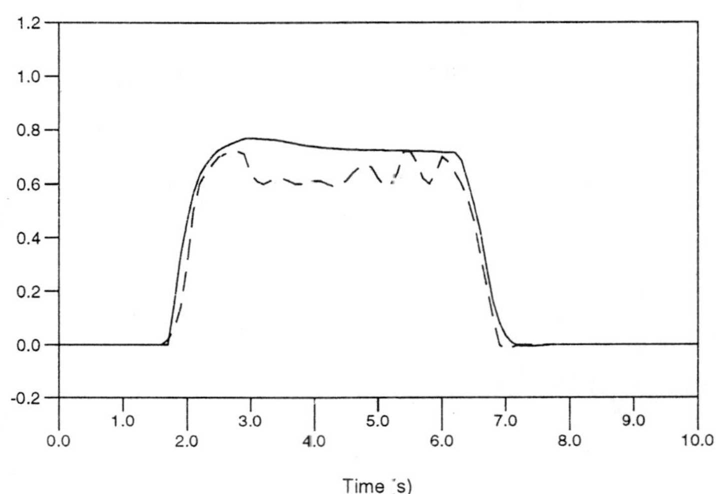

Figure 14 Lateral Acceleration

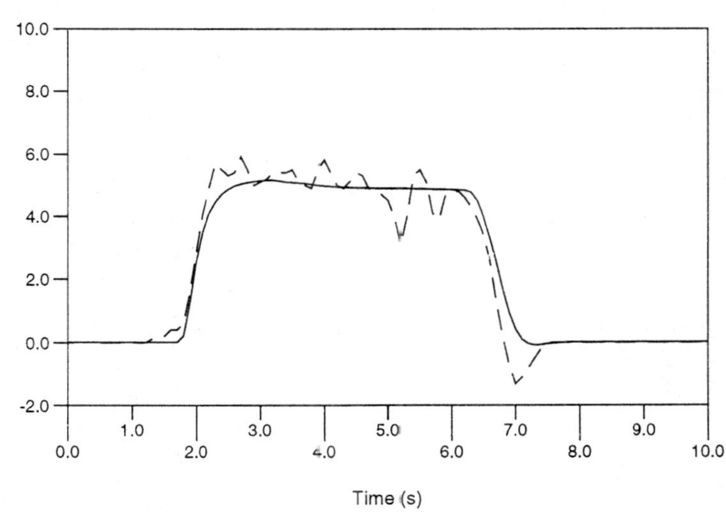

Figure 15 Roll Angle

The essential contribution of rig measurement to suspension design and development

J P WHITEHEAD BA, CEng, MIMechE
MIRA, UK

1.   SYNOPSIS

Laboratory based evaluation is becoming increasingly important during the development of vehicle chassis systems. This paper is concerned with one particular type of laboratory testing, that is the use of a Kinematics and Compliance measurement facility (or K&C Rig) to characterise suspension and steering systems.

After briefly reviewing the history of K&C testing, a description is given of the latest state-of-the-art machine installed at MIRA. The importance of a K&C rig for suspension development and for providing essential input data for computer models is discussed.

2.   INTRODUCTION

Traditionally, the development of vehicle suspension and steering systems, to achieve desired ride and handling performance, has been undertaken by engineers using mainly subjective techniques during road and proving ground testing.

In recent years, the trend has been to move away from this solely qualifying approach to a quantifying one using objective methods. The trend is being accelerated by the increasing use of computer analysis and prediction techniques, which are being used to improve the effectiveness of the design and development processes and to reduce the time from concept to production of a new vehicle.

For a new product, design criteria may be established initially both from experience and in conjunction with benchmarking evaluations of an existing product and of competitor 'target' vehicles. Subsequently, suspension designs are refined with the aid of computer ride and handling models supported by objective testing.

Objective testing takes place on the proving ground, using instrumented vehicles, and in the laboratory. K&C testing is an important facet of the latter, and involves the use of a rig for pseudo-static measurement of the kinematic and compliance (or elasto-kinematic) characteristics of the suspension and steering systems.

Use of such a suspension measurement facility is the only sure way to obtain relevant design information from a competitor vehicle, and to understand how that vehicle achieves its performance. In subsequent development and refining of design, the effectiveness of, and confidence in, the computer modelling approach is enhanced by the use of a rig to measure actual suspension characteristics. The data are used to correlate with the theoretical predictions and hence validate the computer models. The data are used also to correlate the observed behaviour of the vehicle during objective dynamic testing and subjective assessment. During development of prototype vehicles, the facility is invaluable for measuring suspension and steering performance to verify design intent and as an aid for final 'tuning'.

Once a vehicle is in production the effects of 'drift' of manufacturing and assembly tolerances may be monitored by periodic K&C testing of vehicles as they come off the production line.

3.   KINEMATICS AND COMPLIANCE MEASUREMENTS

The way the wheels of a vehicle move relative to its body, in response to vehicle whole body motion of bounce, roll and pitch and in response to forces and moments occurring at the tyre to road interface, is controlled by suspension and steering system characteristics. These wheel movements, in turn, influence the nature of the vehicle whole body motion and some aspects of NVH performance. Although these motions occur at low frequencies, they are dynamic events. Despite this, and with the exception of most NHV aspects, much can be learnt about the motions from steady state or pseudo-static studies. Thus, on the proving ground, the steady state circular test yields a large proportion of handling information about a vehicle. Similarly, in the laboratory, a deep understanding of the relationships between wheel movement and vehicle whole body motion may be obtained by evaluating suspension and steering system characteristics in measurements taken at very low frequency, almost down to DC.

Such measurements may be made with a vehicle suspension Kinematics and Compliance Facility (K&C Rig). The Rig measures kinematic characteristics due to suspension and steering system geometries and elasticities, and compliancies due to suspension springs, anti-roll bars, elastomeric bushes and component deformations.

The essence of the machine is that it subjects the vehicle suspension system to a variety of forces and displacements relative to the vehicle body. It applies the forces and displacements slowly so as not to excite any dynamics emanating, for example, from inertias, dampers or elastomers.

Many parameters may be evaluated, to aid the understanding of a vehicle's performance in terms of ride, impact isolation, handling and steering. The principle parameters are: suspension rates and hysteresis, and lateral and longitudinal compliance for ride; suspension rates, roll stiffness distribution and lateral and longitudinal compliance steer for handling; and torque, hysteresis and compliance characteristics within the steering system.

In many instances, the parameters are affected by flexibilities of the vehicle body and, indeed, in most cases it is important that the effects of these flexibilities be included. Also, for convenience and speed of testing, it is desirable to undertake measurements on complete vehicles rather than on subsystems. For these reasons the K&C Rig is designed for whole vehicle testing. However, this does not preclude the measurement of body deflections, or the use of non-standard vehicle to Rig attachment methods to reduce body flexibility effects.

4.   HISTORY OF K & C MEASUREMENTS

The need for good control of vehicle wheel geometry was known even during the early history of the automobile. However, it was the advent of the radial ply tyre in France in the late 1940's that created an imperative requirement for more precise control of suspension kinematics; early work in this area was done by Citroen in close collaboration with Michelin. The need to introduce longitudinal compliance whilst retaining good wheel control saw Peugeot collaborate with Michelin during the next decade (Reference 1).

As the popularity of the radial tyre grew worldwide, Michelin extended its 'Suspension Static Test Facilities' firstly into the UK and later into the USA. Traditionally these rigs have moved the car against a fixed ground plane, with the wheels supported on air bearings to allow free changes of track, wheelbase and steer. In the roll test the vehicle has not been constrained to roll about a defined axis but has been free to do so about its own roll axis. The Michelin system is powered by a mixture of electro-mechanical and pneumatic actuators.

Probably the most well known machine is the Chevrolet VHF ('Vehicle Handling Facility') designed and built by General Motors in the USA (Reference 2). This became operational in the early 1970's. In this design, the wheel platforms are servo controlled in the ground plane to provide forces and moments if desired, or to provide zero forces and moments to allow for free track, wheelbase and steer changes. During the roll test, the roll axis is imposed. The machine is powered by hydraulic actuators.

Several vehicle manufacturers have been inspired to develop their own facilities based upon the Michelin system or the Chevrolet VHF, or upon some hybrid solution. In addition the VHF has formed the design basis of a number of machines commercially available around the world.

A third line of development, with the title of SPMD ('Suspension Parameter Measurement Device'), started in the early 1960's at Cranfield Institute of Technology in England under the direction of Professor J R Ellis (References 3 and 4). In 1983 the National Highway and Traffic Safety Administration (NHTSA) of the US Federal Department of Transport sponsored an SPMD which was installed at the NHTSA operation in East Liberty, Ohio (References 5, 6 and 7).

The SPMD design was further developed for a machine which was purchased by Goodyear Tire and Rubber Company and installed at its site at Akron, Ohio (References 8 and 9). This machine has been in continuous development since that time, and it received an extensive upgrade in May 1993. A major benefit of this has been a substantial speeding up of test operation. The SPMD moves the vehicle body against a fixed ground plane, and the philosophy of operation resembles that of the GM VHF. As with the latter, the wheel pads are servo controlled (no air bearings). During the roll test, to ensure correct lateral weight transfer across an axle, the vertical height of the car body is adjusted to retain total normal ground load of the axle at its original static value. Actuation is provided by electro-mechanical means.

5.   CHOOSING A FACILITY FOR MIRA

After a lengthy survey of commercially available solutions, MIRA decided that its Kinematics and Compliance Rig should be based on the SPMD concept. The reasons for this decision were as follows:

- The SPMD design had been under development for 30 years.

- At the time of the decision, in its then latest (Goodyear) form, it had been in successful operation for the previous four years.

- The SPMD is highly automated and can produce a complete map of suspension characteristics within an eight-hour day (vehicle attachment or non-routine testing may increase this time).

- The SPMD produces data at the end of each individual test sequence so that results may be viewed as the test programme progresses.

- A number of US vehicle manufacturers have found the Goodyear facility so useful that it has been in continuous heavy usage since installation.

- Using several vehicles in comparative testing with the Chevrolet VHF, identical results have been obtained in the areas of VHF competence; however, the SPMD develops data which are not available from the VHF.

- The SPMD concept is based on the use of proprietary actuators and transducers, and the software control and analysis routines are constructed around proprietary packages; only the mechanical assembly is purpose built. The concept of using proprietary equipment eases the tasks of calibration and maintenance.

- The SPMD does not employ high pressure hydraulic systems.

- The SPMD offers a cost effective solution available now.

5.1 Supplier of the Facility

The current Goodyear SPMD is a two wheel station device that requires the test vehicle to be measured one axle at a time. From the outset it was envisaged that the MIRA K&C Rig would be a four wheel station machine enabling two-axle operation and, hence, a further reduction in a test time from that available currently on the Goodyear facility. It was also envisaged that the Rig would be capable of accommodating all vehicles from the smallest passenger cars to the largest panel vans.

To achieve efficient operation and to ensure correct roll stiffness data from the roll test, it was necessary to introduce a pitch capability to the design for the MIRA machine. During the roll test, the Rig must be capable of simultaneous adjustment of pitch angle and vertical height of the car body, thus enabling the total vertical load at each axle to be maintained constant.

These changes in test control requirements demanded a radical re-design of the mechanical arrangements. With this in mind, it was decided to seek a builder for the Rig in the UK. The company chosen was Anthony Best Dynamics Limited (ABD) of Bradford-on-Avon, England. ABD has many man-years of experience in design and development of vehicle suspension systems. This has involved corresponding experience in the acquisition of kinematic and compliance information. Additionally ABD has a history of designing, developing and marketing a range of special purpose PC-based test and measurement systems for use in noise and vibration work.

6. THE MIRA K & C RIG

6.1 Facility Description

The facility, depicted in Figure 1, comprises five sub-systems:

- The bounce/roll/pitch frame, to which the vehicle is clamped

- The steering input machine and instrumentation

- The wheel pads and instrumentation

- Signal processing and control hardware

- Signal processing and control software

Although the Rig is complete and operational, Figure 1 (an artist's impression) more clearly conveys an appreciation of the rig and its mechanical elements than does a photograph.

Bounce/Roll/Pitch Frame

The standard methods of clamping the vehicle to the frame is by the use of pinch weld clamps.

Kinematics testing is achieved by moving the frame (and hence the vehicle body) in bounce, roll and pitch. The frame is moved by six electrically driven, fine pitch screw jacks under servo control by the control system. Three of the jacks are disposed vertically, two longitudinally and one laterally.

Alternative methods are available for clamping vehicles which do not have pinch welds under the sills.

Transducers are installed to measure the bounce, roll and pitch motions of the bounce/roll/pitch frame.

Steering Input Machine and Instrumentation

The Steering handwheel of the test vehicle may be replaced by an instrumented steering input machine. This is capable of providing inputs at defined torque levels and angular displacements, to defined amplitudes at defined angular velocities. The installed instrumentation will measure the steering torques, angles and velocities applied.

Wheel Pads and Instrumentation

The heart of the K&C Rig comprises the wheel pads (or platforms) and associated instrumentation.

Compliance testing of steering and suspension is achieved by movements of the wheel pads. Each pad can rotate freely or be motored about a vertical (z) axis and can be motored in longitudinal (x) and lateral (y) directions. Within each pad are Kistler load cells to measure three orthogonal forces and three orthogonal moments in the tyre to pad contact area.

The control system allows the wheel pads to be servo-controlled to zero forces and moments to allow them to follow vehicle roadwheel motion during tests involving movement of the vehicle body. Alternatively, when the body is held fixed, the pads can be used to apply forces and moments to the vehicle roadwheels.

Around each wheel pad is a framework (not shown in Figure 1) supporting the other measurement instrumentation. Each framework supports five optical encoders. These are rotational 'yo-yo' devices that are connected to the wheel under test by long, fine wires. A separate transducer monitors castor angle change of the suspension.

Analysis of the signal outputs of the encoders and castor transducer yields the translational and rotational data required for each vehicle wheel.

Signal Processing & Control Hardware

All control and data analysis is serviced by an IBM compatible high performance PC. The Kistler load cells are connected to Kistler charge amplifiers, the outputs of which are fed to the PC via a 16-bit analogue to digital converter. All other transducers are digital, and their signals are fed directly to data acquisition cards within the PC. The system provides for 16 channels of additional transducers.

Signal Processing & Control Software

The system software is custom written by ABD Ltd. using Magic, PMAC and Dia-PC.

The control system is presented to the user in a menu format so that a suite of tests may be defined, and, within the suite, the control strategy for each test may be defined. Upon command the system will then execute the suite of tests.

The format for presentation of each test type may be defined by the user. The results of each test are presented at the end of that particular test sequence.

6.2   Range of Tests

The following lists the range of tests included in a 'standard' package.

Vertical Test

- Brakes are applied.

- Steering Handwheel is fixed.

- The vehicle body is cycled vertically in bounce.

- Horizontal forces and moments in the tyre contact patches are controlled to zero.

- Bounce and rebound limits and cycle time are specified.

Roll Test

- Brakes are applied.

- Steering Handwheel is fixed.

- The vehicle body is subjected to roll motion. This may be at fixed height about a fixed axis, or combined with bounce and pitch motion under servo control to maintain constant total vertical load on each axle.

- Horizontal forces and moments in the tyre contact patched are controlled to zero.

- Roll angle limits and cycle time are specified.

Steering System Tests

- Brakes are free.

- Handwheel steering input is applied cyclically.

- Horizontal forces and moments in the tyre contact patches may be controlled to zero or to some other values.

- Steered roadwheels may be locked in position.

- Steer angle, or torque, limits and cycle time are specified.

Lateral Compliance Tests

- Brakes are free.

- Steering handwheel is fixed.

- The wheel pads are cycled in lateral motion in phase or in opposition.

- Longitudinal forces in the tyre contact patches are controlled to zero.

- Steer aligning torques in the tyre contact patches are controlled to zero or are controlled in relation to the lateral forces.

- Lateral force limits and cycle time are specified.

Longitudinal Compliance Tests

- Brakes are applied normally for testing only in the forward braking sense.

- Transmission is locked normally for testing only in the forward acceleration sense.

- The wheel pads are cycled in longitudinal motion.

- Lateral forces and steering aligning torques in the tyre contact patches are controlled to zero.

- Longitudinal force limits and cycle times are specified.

6.3   Test Results

The format for presentation of results is specified by the operator. Results are available at
the end of each test sequence on PC screen display. Printed results are available immediately
upon completion of a test programme.

MIRA offers a 'standard' package of tests and result parameters as listed below,
although it is acknowledged that many clients will have custom requirements, particularly
with regard to presentation of results. MIRA's facility and service is flexible to cater for
these requirements.

Vertical Test

- Wheel Vertical Rate (contact patch vertical force versus wheel centre vertical
  displacement).

- Tyre Radial Rate (contact patch vertical force versus tyre vertical compression
  displacement).

- Bump Steer (roadwheel steer change versus wheel centre vertical displacement),
  straight ahead and at small nominal steer angles.

- Bump Camber (roadwheel camber change versus wheel centre vertical displacement).

- Bump Castor (roadwheel castor rotation versus wheel centre vertical displacement).

- Track Change (contact patch lateral displacement versus wheel centre vertical
  displacement).

- Wheelbase Change (contact patch longitudinal displacement versus wheel centre
  vertical displacement).

Roll Test

- Roll Stiffness (roll moment versus roll angle).

- Wheel loads (roadwheel vertical loads versus roll angle).

- Roll Stiffness Distribution (ratio of individual axle roll stiffness to total vehicle roll
  stiffness).

- Roll Steer (roadwheel steer change versus roll angle).

- Axle Steer (steer change of beam axle versus roll angle).

- Roll centre height of each 'axle'.

Steering System Tests

- Roadwheel Steer versus Handwheel Steer.

- Ackermann curves.

- King Pin offset, scrub radius and mechanical trail.

- Handwheel Steer versus Handwheel Torque for various control strategies of Roadwheel Steer.

- Steering System stiffness.

Lateral Compliance Tests

- Lateral Compliance (roadwheel centre lateral displacement versus contact patch lateral force).

- Lateral Compliance Steer (roadwheel steer change versus contact patch lateral force); this may be a family of curves with different effective longitudinal moment arms acting in the contact patch.

- Camber Lateral Compliance (roadwheel camber angle change versus contact patch lateral force).

Longitudinal Compliance Tests

- Longitudinal Compliance (roadwheel centre longitudinal displacement versus contact patch longitudinal force).

- Longitudinal Compliance Steer (roadwheel steer change versus contact patch longitudinal force).

- Castor Longitudinal Compliance (roadwheel castor angle change versus contact patch longitudinal force).

6.4   Facility Specification

Dimensions

Number of Axles (Number of Wheel Pads)	2(4)
Vehicle Nominal Maximum Kerb Mass (kg)	4000
Wheelbase Range (mm)	2025-3980
Track Range (mm)	1150-1725

Motions and Measurements

	Range
Bounce Movement (mm)	±150
Roll Angle (deg)	±10
Pitch Angle (deg)	±8
Wheel Pad Vertical (z) Load (N)	0 to 30 000
Wheel Pad Longitudinal (x) Load (N)	±20 000
Wheel Pad Lateral (y) Load (N)	±20 000
Wheel Pad Aligning (Mz) Torque (Nm)	±1000

Wheel Pad Rotational Speed (deg/sec)	±5
Wheel Centre Vertical Movement (mm)	±50
Wheel Centre Longitudinal Movement (mm)	±150
Wheel Centre Lateral Movement (mm)	±100
Wheel Steer Angle (deg)	±60
Wheel Camber Angle (deg)	±20
Wheel Rotation (Castor) (deg)	±25
Handwheel Steer Angle (deg) (dual range)	±1000 (±100)
Handwheel Steer Torque (Nm) (dual range)	±100(±10)
Handwheel Rotational Speed (deg/sec)	±90

7. CONCLUSIONS

MIRA's new facility provides an up-to-the-minute, state-of-the-art capability for kinematic and compliance testing of vehicle suspensions. MIRA believes that its new rig is unique in being a four wheel station machine that provides a commercially available service for the complete characterisation of a vehicle's suspension within two days. This time period is expected to reduce as experience is gained in the use of the facility.

8. REFERENCES

1. DIXON, J.C. Noise-reduction Design-features of the Peugeot 404, <u>MIRA Report 1963/2</u>, 1962.

2. NEDLEY, A.L. and WILSON, W.J. A New Laboratory Facility for Measuring Vehicle Parameters Affecting Understeer and Brake Steer, 1972, <u>SAE 720473</u>.

3. ELLIS, J.R. A Study of Suspension Mechanisms, <u>Cranfield Institute of Technology, ASAE Report No.5</u>, 1969.

4. ELLIS, J.R. and BUTLER, D.M. Analysis and Measurement of the Relative Motions Between the Road Wheels of a Vehicle and the Road Surface, <u>Proceedings of the Institution of Mechanical Engineers</u>, Vol 186 57/72.

5. GARROT, W.R. et al. VRTC's Light Vehicle Stability and Control Research Program, Report No 1: The Suspension Parameter measurement Device, <u>NTIS</u>, 1986.

6. ELLIS, J.R. et al. The Design of a Suspension Parameter Measurement Device, <u>SAE 870576</u>, 1987.

7. BELL, S.C. et al. Suspension Testing Using the Suspension Parameter Device, <u>SAE 870577</u>, 1987.

8. COOVERT, D.A. et al. Design and Operation of a New Type Suspension Parameter Measurement Device, <u>SAE 920048</u>, 1992.

9. CHEN, H.F. et al. Suspension Testing on the New-Type Suspension Parameter Measurement Device, <u>SAE 920049</u>, 1992.

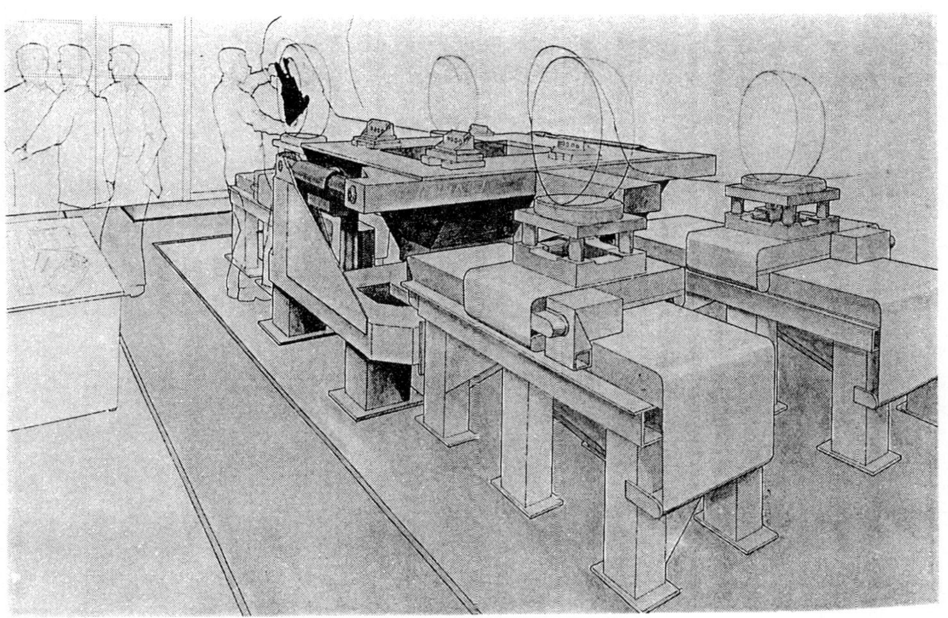

FIG 1   MIRA's Vehicle Suspension Kinematics and Compliance Facility
R20666

igital control for vehicle active suspension systems

M EL-DEMERDASH
Iwan University, Egypt
R PLUMMER and D A CROLLA
iversity of Leeds, UK

SYNOPSIS

This paper describes the design of a digital controller for a hydro-pneumatic slow-active suspension system with road preview. Simulation results show that with 100 Hz sampling frequency it is possible to obtain similar performance to that of an analogue controller. The controller is tested on a quarter-car suspension test rig, and experimental and theoretical results are compared.

NOTATION

f	frequency, Hz
K_1, K_2, K_3	feedback gains
K_P, K_I, K_D	PID controller proportional, integral and derivative gains
T_s	sampling time, s
x_b	sprung mass vertical displacement, m
x_w	wheel vertical displacement, m
x_{o1}	road input displacement at wheel position, m
x_{o2}	road input displacement at preview position, m

1 INTRODUCTION

In recent years significant progress has been made in the analysis and design of digital control systems. They overcome the disadvantages associated with analogue controllers, in which the elements are usually hard-wired, making it difficult to make design changes. Also component ageing and sensitivity to environmental changes can cause problems. Most previous work on active suspension control systems [e.g. 1-5] assumes that the feedback and feedforward loops of the control structures are of a continuous analogue type.

Fig. (1) shows a quarter car model incorporating a hydro-pneumatic slow-active suspension system with its analogue controller structure [5]. This controller is used to control the suspension system vibration within a bandwidth of 6 Hz, and can also adjust the suspension ride height to meet payload changes (self levelling). In [6], this controller is used to study the

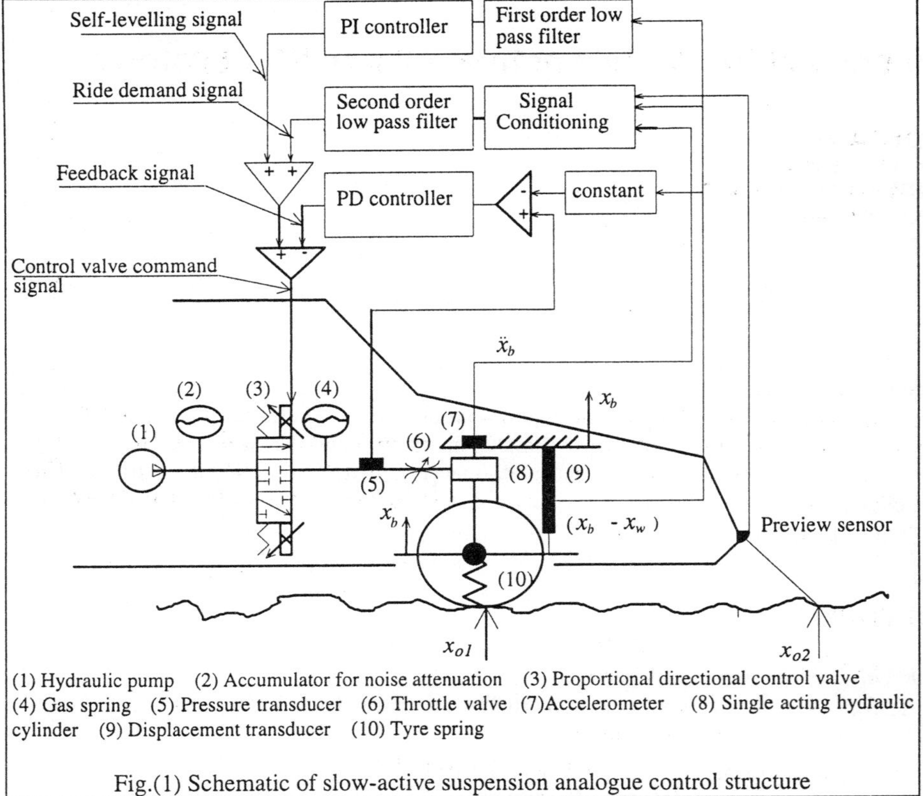

(1) Hydraulic pump   (2) Accumulator for noise attenuation   (3) Proportional directional control valve
(4) Gas spring   (5) Pressure transducer   (6) Throttle valve   (7) Accelerometer   (8) Single acting hydraulic
cylinder   (9) Displacement transducer   (10) Tyre spring

Fig.(1) Schematic of slow-active suspension analogue control structure

suspension system performance with and without preview control including the effects of component non-linearities in the system. In this paper, a digital controller equivalent to that given in Fig. (1) is designed and a suitable sampling time is selected in order to implement the controller for a hydro-pneumatic slow-active suspension test rig.

2 DESIGN OF A PRACTICAL DIGITAL CONTROLLER

The closed loop digital control scheme for the slow-active suspension is shown in Fig (2). In this system, the measured variables are body acceleration, suspension working space (SWS), and gas spring pressure. It is also assumed that the road height at the wheel and in front of the wheel (preview) can be measured.   These signals are fed to the controller after signal conditioning in the form of differentiation, integration and low pass filtering as required.

There are three parts to the controller:
- Ride control with controller gains K_0, K_1, K_2 and K_3. Setting K_0 to zero is equivalent to having no preview. These gains can be calculated using optimal control methods [5].
- Proportional plus derivative (PD) servo-loop to ensure accurate gas spring pressure demand tracking.
- Proportional plus integral (PI) ride height controller.

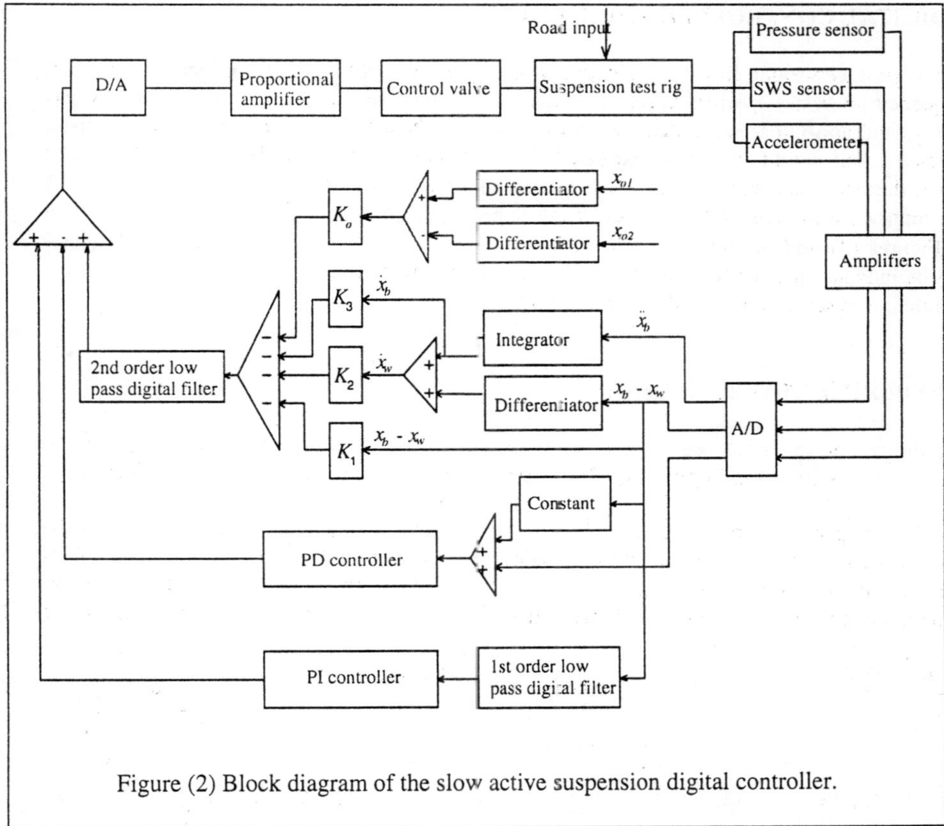

Figure (2) Block diagram of the slow active suspension digital controller.

The implementation of each part of the controller is represented in Table 1 by a z-domain transfer function. Each has been converted from the continuous time equivalent either by the backward trapezoidal method, or by bilinear transformation.

Table 1 Discrete-time implementation for controller elements.

Differentiator	$\dfrac{1-z^{-1}}{T_s}$	1st order low pass filter:	$\dfrac{a_o}{b_o+b_1z^{-1}}$
Integrator:	$\dfrac{T_s}{2}\left(\dfrac{1+z^{-1}}{1-z^{-1}}\right)$	2nd order low pass filter:	$\dfrac{a_o+a_1z^{-1}+a_2z^{-2}}{b_o+b_1z^{-1}+b_2z^{-2}}$
PD controller:	$\dfrac{a_o+a_1z^{-1}+a_2z^{-2}}{1+b_1z^{-1}}$ where $a_o=K_P+\dfrac{K_D}{T_s},a_1=-K_P-K_D\dfrac{2}{T_s}, a_2=\dfrac{K_D}{T_s}, b_1=-1$		
PI controller:	$\dfrac{a_o+a_1z^{-1}+a_2z^{-2}}{1+b_1z^{-1}}$ where $a_o=K_P+K_I\dfrac{T_s}{2}, a_1=-K_P+K_I\dfrac{T_s}{2}, a_2=0, b_1=-1$		

3 DIGITAL CONTROLLER SIMULATION

The digital controller has been simulated in order to investigate the effect of sampling frequency on system performance. The non-linear model [6] that incorporates adiabatic gas springs and non-linear valve characteristics is used for the simulation to represent the realistic hardware that would in practice be used with the digital controller. Using a random road input, the body acceleration and valve command signal have been compared with those for the continuous system for two values of the sampling frequency, $1/T_s$. The results are shown in Figures (3) and (4) for a vehicle travelling at 20m/s on a simulated major road. These results indicate that the body acceleration and valve command signal are very similar to the continuous system for the 100Hz sampling frequency.

4 EXPERIMENTAL RESULTS

The digital controller has been tested using a quarter car rig. The layout of the test rig with its control loop is shown in Figure (5). Some parameter values for the rig are given in Table 2.

The levelling control performance is tested by injecting a step input signal of 6 V to the control valve. This is equivalent to a step input of external body force of 800 N. A comparison between the levelling control results for this system and those for the passive system are shown in Figure (6). It is clear that there is some agreement between the measured

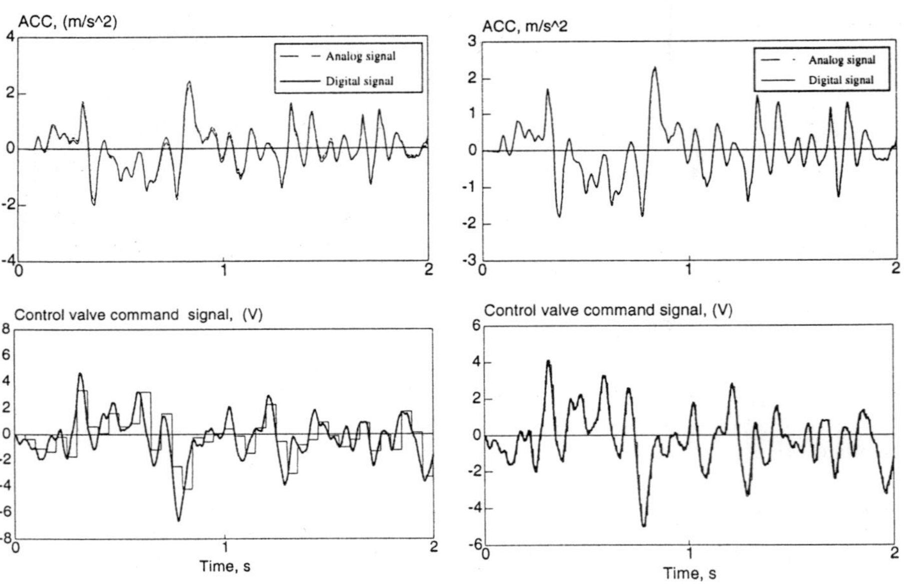

Fig. (3) Digital controller performance for non-linear slow-active system with preview at sampling frequency 20 Hz. The preview time is 0.075 s.

Fig. (4) Digital controller performance for non-linear slow-active system with preview at sampling frequency 100 Hz. The preview time is 0.075 sec.

and simulated results, but the experimental results exhibit a stepwise movement which is due to friction in the hydraulic cylinder.

The performance was measured for the slow-active suspension with and without preview control using the proposed controller, and compared to an equivalent passive system. The input was representative of a typical major road. The power spectral density results for body acceleration, dynamic tyre load and suspension working space are shown in Figure (7). It can be seen from these results that the proposed controller works well and there are improvements in the ride comfort for the slow-active with and without preview over that passive system.

Root-mean-square (r.m.s) values for both the measured and simulated results are compared in Fig. (8). There is a 36% improvement in the r.m.s value for body acceleration and 6% improvement in dynamic tyre load for the system with preview over the passive system. The suspension working space stays approximately constant. However, agreement between

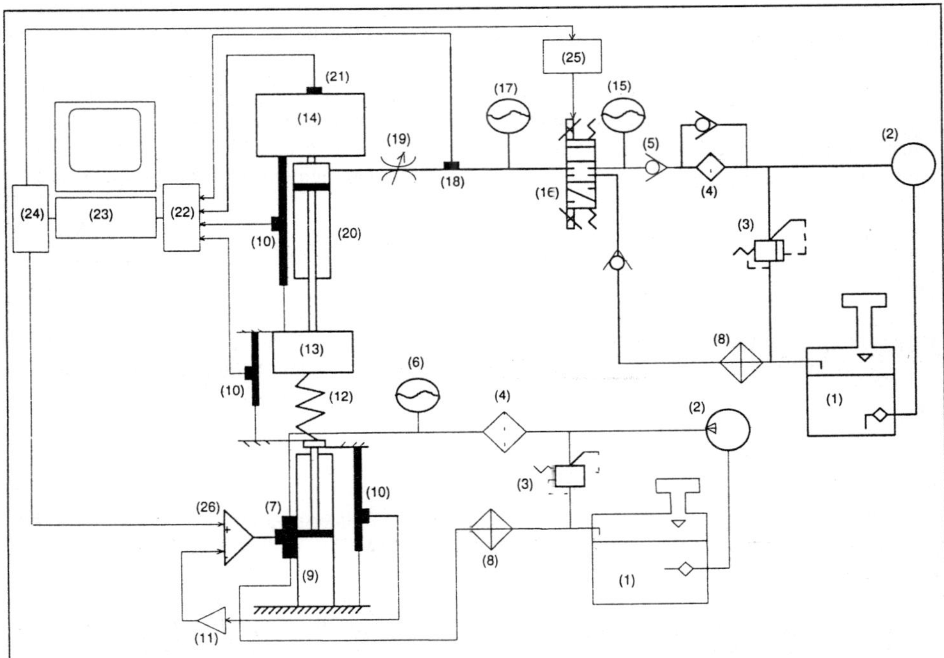

(1) Oil tank   (2) Hydraulic pump   (3) By-pass valve   (4) Hydraulic filter   (5) Check valve   (6) Input exciter accumulator   (7) Servo valve   (8) Cooler   (9) Double acting hydraulic cylinder   (10) Displacement transducer   (11) Transducer amplifier   (12) Tyre spring   (13) Unsprung mass   (14) Sprung mass   (15) Main accumulator   (16) Proportional control valve   (17) Gas spring   (18) Pressure transducer   (19) Throttle valve   (20) Single acting hydraulic cylinder   (21) Accelerometer   (22) Analogue to digital converter   (23) 486 personal computer   (24) Digital to analogue converter   (25) Proportional amplifier.

Fig. (5) Layout for the hydro-pneumatic slow-active suspension test rig.

Table 1 Test rig component specifications

Component	Specifications
Sprung mass	322 kg
Unsprung mass	71 kg
Tyre spring stiffness	120000 N/m
Gas spring	Pre-charge pressure 1.4 MPa and pre-charge volume 0.5 litre
Single acting cylinder	32 mm diameter and ±100 mm stroke

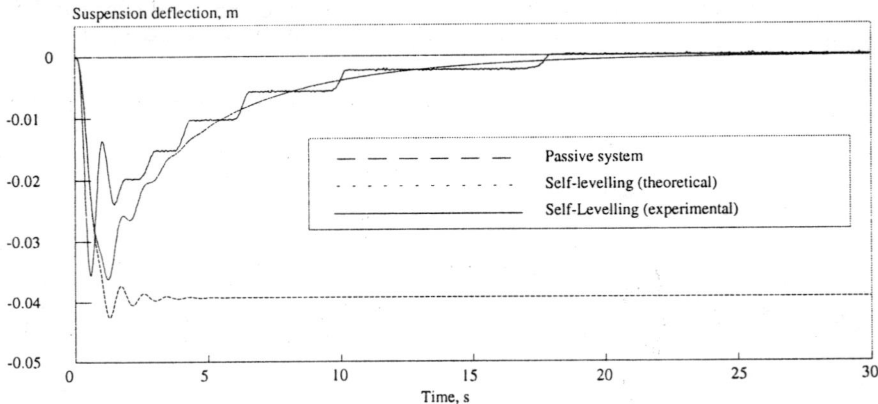

Fig. (6) Experimental and theoretical self-levelling performance compared to passive system. Experimental is for injected signal of 6V and theoretical is for body step input of 80 Kg

measured and simulated results is not exact; the measured performance is actually better than that predicted by the theoretical simulation. This discrepancy clearly requires further consideration in the on going research programme. However, it does not detract from the main issues which are:

- that the digital controller works satisfactory
- that the results with the hydro-pneumatic slow-active systems reveal the anticipated benefits.

5 CONCLUDING REMARKS

The selection of sampling rate is one of the most important decisions to be made in the design of digital controllers. It should be fast enough to avoid aliasing and to achieve the required performance of the final control system. The recommended value should be at least 5 times the desired closed loop bandwidth [7]. The simulation results for the hydro-pneumatic slow active suspension showed that a 100 Hz sampling rate would be suitable in this case.

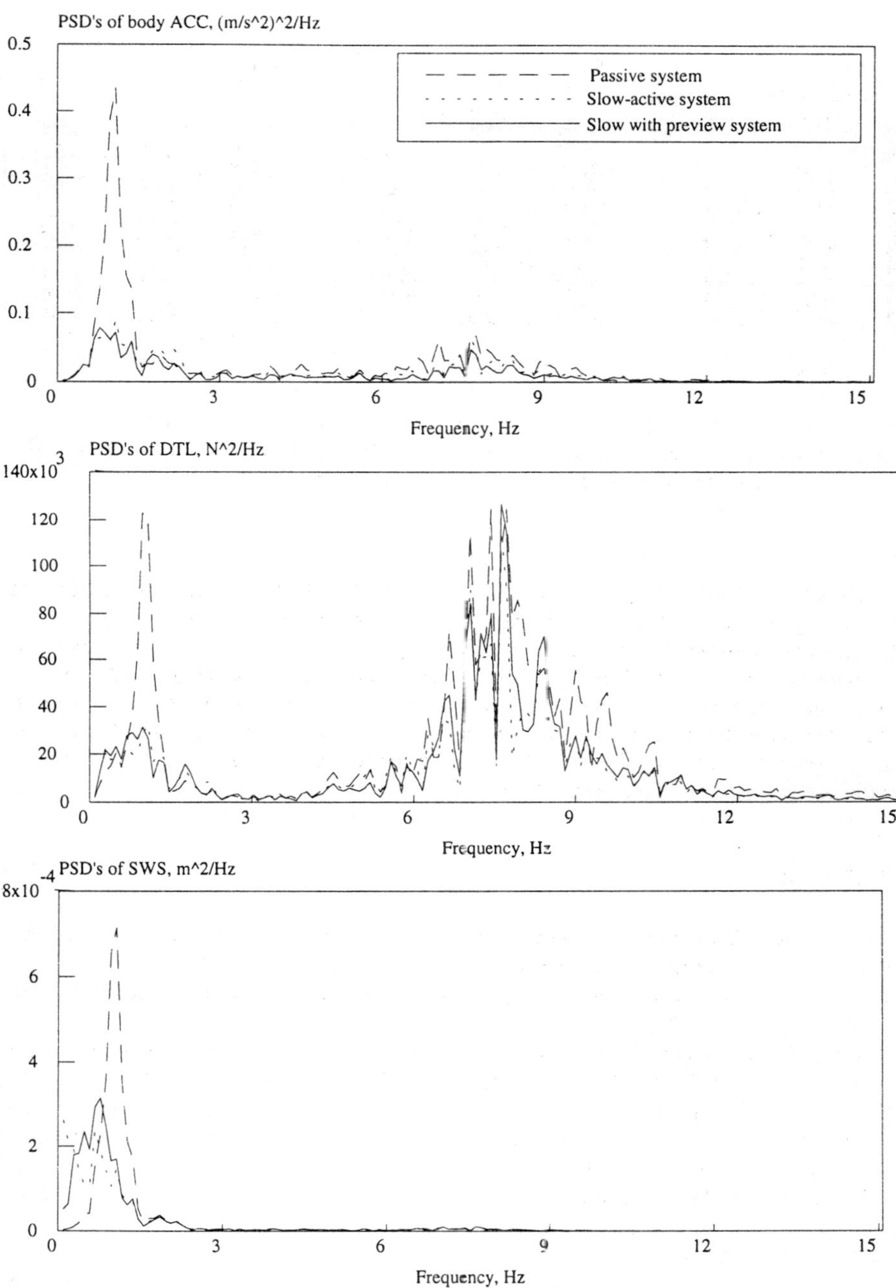

Fig. (7) Comparisons of the experimental results for body acceleration, dynamic tyre load and suspension working space power spectral densities for different suspensions on a major road with vehicle speed of 20 m/s. The preview time is 0.075 s (1.5 m at 20 m/s)

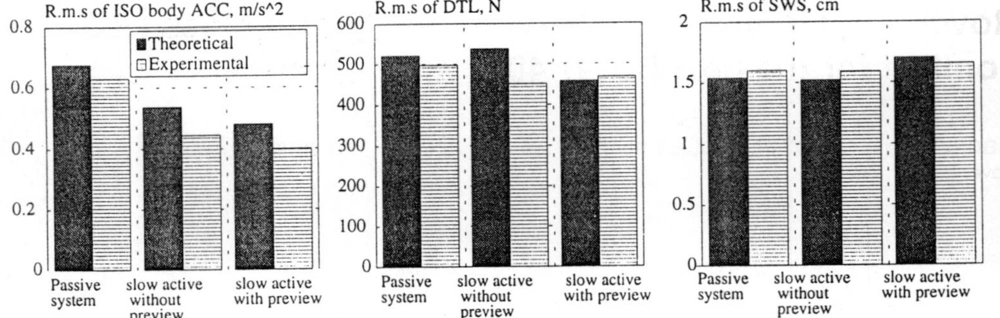

Fig (8) Comparison of experimental and theoretical r.m.s. values for body acceleration, suspension working space, and dynamic tyre load for different suspensions on a major road with vehicle speed of 20 m/s. The preview time is 0.075 s.

A digital controller has been designed and implemented which includes signal estimation and signal filtering as well as PI and PD digital controllers. A computer program, written in C, implements the controller in real time to control a hydro-pneumatic slow active suspension system with preview control. The digital controller is shown to work satisfactorily and the resulting improvements in suspension performance measured on the experimental rig were in line with the benefits predicted by the theoretical simulations. The ride control and self-levelling algorithms are both shown to function correctly together, although friction in the hydraulic actuator remains a problem.

REFERENCES

(1) Bender E. K. " Optimum linear preview control with application to vehicle suspension", Trans. ASME Journal of Basic Engineering , Vol.90, No. 2, 1968, pp.213-221.

(2) Hac, A., "Optimal linear preview control of active vehicle suspension", Vehicle System Dynamics, Vol.21(3), 1992, pp.167-195.

(3) Pilbeam, C. and Sharp, R. S., "On the preview control of limited bandwidth vehicle suspensions", Proc. Ins. Mech. E., Part D, Vol.207, 1993, pp.185-194.

(4) Huisman, R. G. M., Veldpaus, F. E., Voets, H.J.M. and Kok, J. J., " An optimal continuous time control strategy for active suspensions with preview", Vehicle System Dynamics, Vol. 22(1), 1993, pp. 43 -55.

(5) El-Demerdash, S. M. and Crolla, D. A. " Hydro-pneumatic slow-active suspension with preview control", (submitted for publication to Vehicle System Dynamics).

(6) El-Demerdash, S. M. and Crolla, D. A. "Effect of non-linear components on the performance of a hydro-pneumatic slow-active suspension system", (submitted for publication to IMechE Part D)

(7) Golten, J. and Verwer, A., " Control system design and simulation", McGRAW-Hill Book Company, 1990.

Rover's system approach to achieving first class ride comfort for the new Rover 400

R BOSWORTH BSc, **J TRINICK** BSc, **T SMITH** BTech, AMIMechE, and **S HORSWILL** BTech
Rover Group Limited, UK

SYNOPSIS

In order to achieve Rover marque values for the ride comfort of a new medium sized car, a new approach to ride development was evolved. A multifunction team was formed to adopt a system approach to develop the vehicle. The engineering of the suspension and engine mount systems was carried out as an integrated exercise.
A PC based ride model was created to assist in the design and development of the spring/damper and engine mount settings. A new design of suspension damper valving incorporating linear one to one damper characteristics, and a decoupled hydrabush engine mount, resulted in a medium sized vehicle with a first class ride.

NOTATION

f = frequency Hz
f_0 = frequency of maximum out-of-phase stiffness Hz
A = in-phase stiffness at f_0 N/mm
B_1 = maximum deviation of in-phase stiffness N/mm
B_2 = minimum deviation of in-phase stiffness N/mm
C_1, C_2 = shape factors
D_1, D_2 = bandwidth factors
F = maximum out-of-phase stiffness N/mm
G_1, G_2 = shape factors
H_1, H_2 = bandwidth factors

1.0 INTRODUCTION

To achieve Rover marque values and the project objectives for the ride performance of the new Rover 400, a new approach to ride was required. A ride team drawn from design analysis and design and development engineers from chassis suspension and engine mounting systems was established. The team's task was to set new standards for ride comfort for a medium sized car without adversely affecting the handling criteria.

The difficulty of obtaining an overall good riding vehicle with passive suspension and elastomeric power unit mountings has long been recognised. The primary ride (0 - 6 Hz) is generally controlled via the setting and tuning of the front and rear suspension springs, dampers and bushes. The spring rates are set to produce the rigid body bounce and pitch modes of vibration in the range 1 to 1.5 Hz whilst the levels of damping are such as to give good body control. If the dampers are set too soft, to obtain good secondary ride (in terms of good power unit control), then there is an adverse effect on primary ride and handling (noticeably lane change type manoeuvres). Wheel/tyre patter may also arise. Too soft a ride also generates 'travel' sickness, and therefore the emphasis of the suspension characteristics is on primary rather than secondary comfort.

Secondary ride (6 - 30 Hz) control has to be achieved, therefore, through the power unit mounts. With conventional elastomeric mounts the level of damping that can be obtained, avoiding creep type phenomena and durability problems, is insufficient to give the high degree of control required in today's vehicles. Raising the frequencies of the power unit modes to be outside the wheel excited range is not a viable proposition with vertically installed four cylinder power units because of the intolerable noise levels that would result. What is required is a mount that has a high level of damping at the frequencies of the main power unit modes and very little damping away from these frequencies. Such a mount is the hydraulically damped engine mount (hydramount or hydrabush). Ideally, for engine forcing isolation, the mount stiffness should not increase greatly with increasing frequency of excitation. It is possible to achieve stiffness characteristics that have values lower than a conventional low damped elastomer type mount at high frequencies (1).

2.0 MODELLING

Elastomeric mounts are commonly described by their dynamic stiffness properties in the frequency domain. The characteristics are measured at some appropriate amplitude and a nominal value utilised in the model.

The characteristics of hydramounts are quite different from elastomeric mounts and various attempts have been made to represent their behaviour by means of a lumped parameter mass, spring and damper arrangement (2). The parametric values are obtained by curve fitting measured data. The disadvantage with this approach is the need to extend the vehicle model to include the mathematical representation of the hydramount(s). Additionally, it is currently not possible to relate the parameters to the physical properties of the hydramount

The approach taken here was to represent the hydramount characteristics by means of two formulae, one to describe the in phase and the other the out of phase stiffness versus frequency characteristics (3).

An assessment on ride was made through a linear mathematical model used to describe the vehicle performance on a four poster test rig. Such a model has proved it's worth when applied at the concept or early design/development stages of the vehicle life

	Primary	Secondary	Overall
Prototype 400	3.32	2.66	4.2
Production 400	2.99	2.06	3.62
214	3.00	2.80	4.03

Table 1

5.0 DISCUSSION

The feel when driving the new Rover 400 is that of being in a larger, executive class car. The most striking feature is the comfort level, with the absence of sharp jerks over pot holes and road joints and the freedom from a constant fidgeting when travelling over apparently smooth roads.

The primary ride offers a first class ride comfort by gliding over small amplitude long wavelength road undulations, yet maintaining progressive body control over higher amplitude undulations. This eliminates 'crash-through' and 'topping-out' typically associated with conventional damping.

6.0 CONCLUSIONS

A PC based ride model and the team approach to developing ride has enabled:

- All the project objectives to be met.
- A six month reduction in ride development time.
- A first class ride comfort to be achieved in a medium sized car.

REFERENCES

(1)	WEST, J. 'Hydraulically damped engine mounting'. Automotive Engineer.

(2)	MULLER M., WELTIN U., FREUDENBERG C. 'Schwingungskompensation mit aktiven Motorlagern'. 14th International Vienna Engine Symposium, May 1993.

(3)	BOSWORTH, R. FURSDON, P.M.T. 'A Frequency Domain Representation of Hydramounts for Ride Prediction'. SIA 9506C08, 5th EAEC Congress, Strasbourg, 1995.

(4)	ASHLEY, C. 'The Constant Velocity Approach to Vehicle Ride Testing in the Laboratory'. Paper H4, Belgrade International Automobile Engineering Symposium, 1972.

(5)	LUCAS, K., KAFETZIS, G.C. 'Use of a four channel road simulator to optimise the ride quality of a luxury saloon vehicle'. C466/005/93, 1993.

(6)	LEATHERWOOD, J.D., BAKER, L.M. 'A user oriented and computerized model for estimating vehicle quality'. NASA Technical Paper - 2299, 1984.

2.1 Linear Ride Model

Ride investigations in the laboratory are generally performed by either or both a modal and frequency response survey. For a modal study the vehicle would be excited randomly through a small electrodynamic shaker and the response measured at a number of locations. Modal characteristics and natural frequencies are then extracted from frequency response functions. For road and shock type inputs the vehicle would be mounted on a four poster rig and excited with electrohydraulic shakers. Data would be analysed at appropriate points of interest. Initially the frequency response is studied to determine the levels and frequencies of vibration. Although road inputs are mainly random in nature it is possible, indeed in the past the only way, to identify problem frequencies and their relative amplitude by means of a swept sine input applied vertically through the tyres. Such an approach has a lot to offer. Ashley (4) has shown that road inputs can be reasonably well represented by means of a constant velocity swept sine input. Acceptable levels of vibration can be determined from subjective assessment of the vehicle.

Traditionally damper settings in rebound and bump are set with a ratio of 3 to 1 or even higher with force versus velocity characteristics showing a distinct 'knee point', see Fig 1. There appears to be little evidence from an analytical viewpoint for rebound and bump characteristics to be different. Also from wheel velocity measurements taken on vehicles over various 'ride' routes there appeared to be little justification for the damper force to be reduced at high velocity levels i.e. why not have linear damping ?

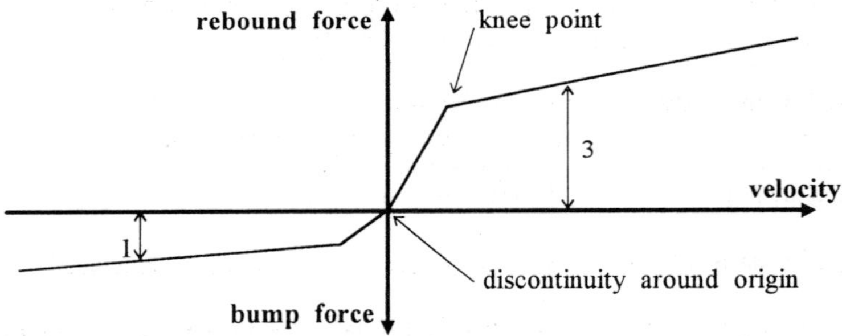

Fig 1 Traditional damper force/velocity curve.

It is with the above in mind that a linear ride model was developed. The main aspects of ride, in terms of occupant comfort, are the vertical, roll and pitch motion of the vehicle body (i.e. 3 degrees of freedom). Allocating the power unit the same degrees of freedom, then with the four wheels, a ten degree of freedom model should give a good first insight into a vehicle's ride.

A decision was made to generate the model on a PC as this was easily accessible to both design and development engineers. Engineers are quite familiar with PC's and especially in the usage of spreadsheets. Therefore a model was created in the Microsoft spreadsheet software Excel. Input to the model is simply performed by typing in the mass, inertia, stiffness properties and geometrical locations into appropriate cells in the spreadsheet. Output is in the form of acceleration versus frequency plots at selected locations on the body.

By comparison with previous vehicle test results it has been shown that the important areas of the model was its ability to represent the dynamic stiffness characteristics of suspension bushes and power unit mounts in the frequency domain.

To validate the ride model, a correlation exercise was undertaken. The seat rail response from a vehicle on an electrohydraulic four poster rig (5) was used for comparison with the model's predicted response. Fig 2 shows the responses at the right hand seat rail during the beginning of the ride development.

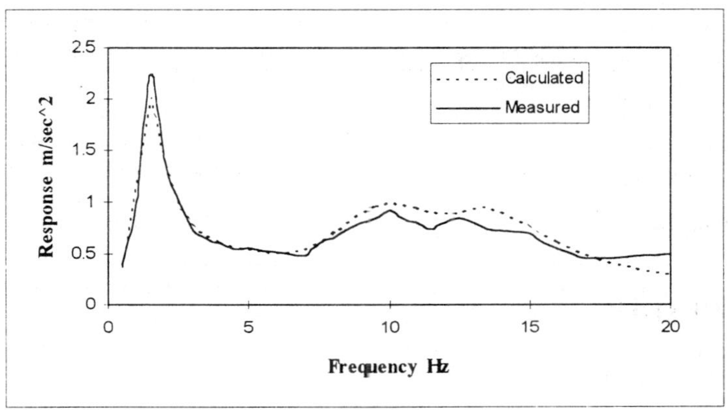

Fig 2 Comparison of actual seat rail response with predicted.

2.2 Hydrabush Curve Fitting

The hydrabush characteristics are represented in the frequency domain by two formulae (3) which describe the in-phase and out of phase stiffness as follows:

The in-phase stiffness is

$$\text{for } f < f_0 \qquad K_i = A + B_1 \sin (C_1 \tan^{-1} (D_1 (f\text{-}f_0))) \qquad (1)$$

and

$$\text{for } f > f_0 \qquad K_i = A + B_2 \sin (C_2 \tan^{-1} (D_2 (f\text{-}f_0))) \qquad (2)$$

The out of phase stiffness is

$$\text{for } f < f_0 \qquad K_o = F \cos (G_1 \tan^{-1} (H_1 (f\text{-}f_0))) \qquad (3)$$

and

$$\text{for } f > f_0 \qquad K_o = F \cos (G_2 \tan^{-1} (H_2 (f\text{-}f_0))) \qquad (4)$$

Fig 3 shows the level of agreement attained using this approach, where the difference between measured and curve fitted results is nowhere greater than 2%.

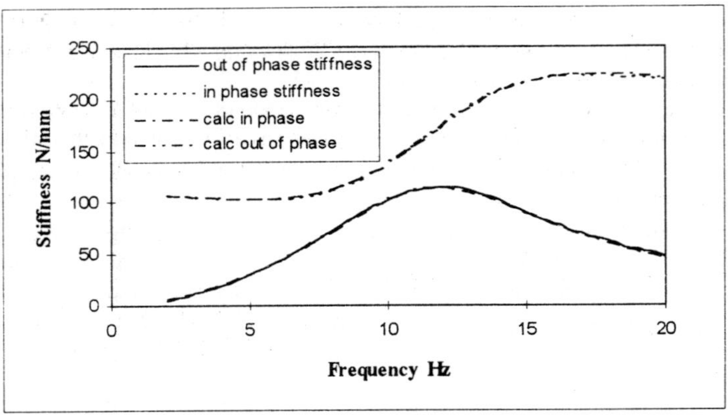

Fig 3 Hydrabush curve fit.

3.0 PROJECT TEAM

3.1 Objectives

The new Rover 400 competes in the Upper Medium sector of the market, where ride comfort is an important dynamic marque value. The car needed to be relaxing to drive, with a distinctive fluent ride, but without compromising driver enjoyment and it had to inspire confidence when driven under all conditions.

At the beginning of the project specific targets were set for the vehicle's dynamic behaviour in the 'Product Design Specification', PDS, document.

The following table shows these targets compared with the Rover 214 existing ratings:

Parameter	Rover 400	Rover 214
Primary Ride	8	7
Secondary Ride	9	6
Engine Shake	9	6
Roll Angle	8	8
Overall Handling	8	7

The ratings range from 1(unacceptable) to 10 (exceptional).

3.2 Team Approach

To meet these stringent targets a new approach to ride was required. The use of conventional chassis technology and development techniques in recent Rover products has resulted in the vehicles being set-up with an emphasis on primary ride control and powertrain isolation. Both these factors compromise the secondary ride of the car. Though the suspension damping is set high, the front and rear damping is rarely optimised to give a balanced ride, the rear being overly stiff relative to the front thereby inducing excessive pitch into the vehicle. Previously the development of engine mounts characteristics were carried out separately to the suspension characteristics and as a result the vehicle was never truly optimised for ride comfort across the frequency range, 0-30 Hz. Learning from these problems on previous models it was realised that the ride of the new Rover 400 would have to be engineered considering all the appropriate components together and not individually. This meant that, when modelling, designing and developing the relevant components, they must be treated as part of one complete vehicle system and not separate sub-systems.

To engineer the ride in this way a 'Ride Team' was formed. The skills make up of the team was intended to open up ride discussion from being the preserve of development engineers. This introduced the analytical skills of Design Analysis and project management from the Design area. The ride team consisted of engineers from; Vehicle Dynamic Analysis, Engine Mount Design and Development, Suspension Design and Development, and Vehicle Simulation Test.

The engineering process that was followed, from analytical modelling through to production release, is shown in Fig 4.

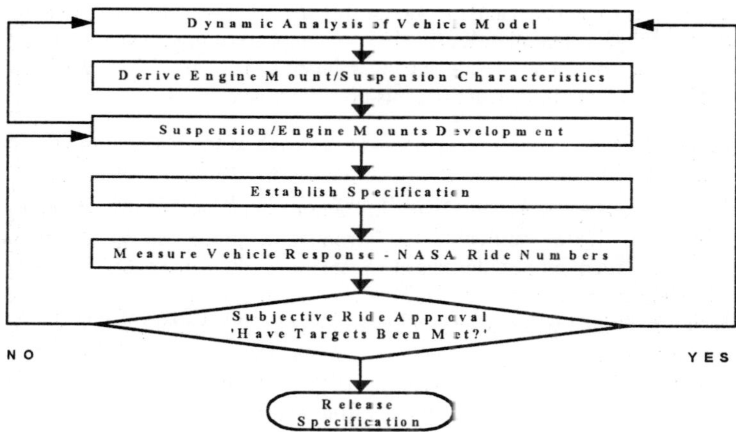

Fig 4. Engineering development process.

This team approach has led to a common direction for the vehicle's character, due to the whole team having the same objectives.

3.3 Timing

The team approach to ride work and the use of the PC based ride model to establish the initial damper characteristics have shortened the time to develop the ride of a vehicle. The ride development of the new Rover 400 took 4 weeks over a 12 month period whereas the old Rover 214 (launched in October 1989) took 9 weeks over an 18 month period.

The arrival of the first prototype vehicles in May 1993 brought the challenge of how to develop the new Rover 400 to give class leading ride.

4.0 RIDE DEVELOPMENT

4.1 Vehicle System Ride Development

The preliminary spring and damper settings gave disappointing ride and handling characteristics. However ride assessment of the new 400 showed it to be better than the existing 200, the stiffer body structure benefiting both noise and vibration, whilst the engine mounting system geometry offered a reduction in engine shake compared with that of the 214. The ride character of the vehicle was largely unchanged from that of the 200 and did not meet the aspirations of the PDS or of Vehicle Dynamics aim of a first class ride.
Discussions within the ride team highlighted engineers' awareness of technologies that could be directed at the new 400 to provide a means whereby the ride aspirations could be met.
These were:

- A new PC based vehicle ride model enabled Design Analysis to evolve a new spring and damping strategy that balanced the primary attitude of the car whilst minimising seat rail acceleration.
- Engine mount engineers and the mount supplier Avon-Clevite had shelf engineered a hydraulic engine mounting bush. This not only gave high levels of low frequency damping as available in current hydrabushes, but gave low stiffness at high frequencies similar to conventional elastomeric bushes.

4.2 Damping Strategy

The initial damping strategy was based on the conventional asymmetric damping, with a rebound to compression force ratio of 3:1. Introduction of the radical one to one linear damping strategy and road springs optimised to minimise pitch imbalance, brought about a significant improvement in the vehicle ride. It being found that large improvements in secondary ride were attained without loss of body control. This benefit was obtained through the use of increased damping at high velocity in bump and rebound and reduced low velocity damping in rebound. This enabled further spring refinements to be made to meet PDS targets of roll and laden handling, allowing stiffer springs front and rear.
The final ride specification produced a vehicle with good secondary refinement, whilst maintaining body control under all conditions. Fig 5 shows how well the final damper settings matched the initial predicted line. A benefit of using the PC based ride model, was that it gave a good base from which to start the ride development. Previous ride work has taken 20-30 iterations to achieve a satisfactory outcome. With the PC based ride model only 16 iterations were required. The high damping forces at high velocities required for the new damping strategy could not be achieved with the existing damper. To meet the requirements, Delphi Chassis Systems, the Rover damper supplier, had to develop a unique valve/piston for the new Rover 400.

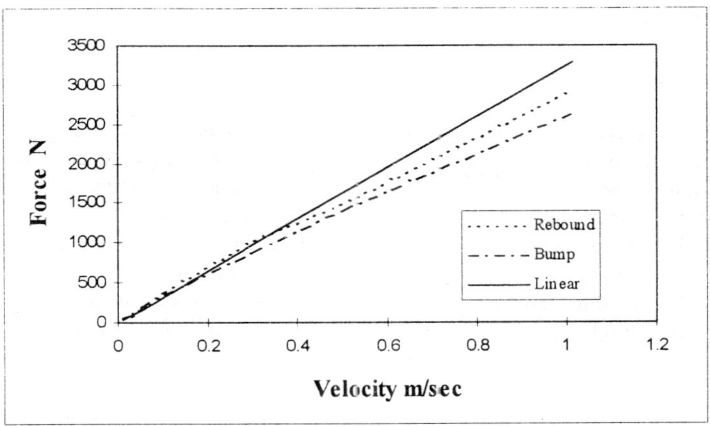

Fig 5. Actual versus Specified Damper Curve.

4.3 Engine Mount Development

Further improvements to secondary ride required better control of engine shake. To achieve this high damped elastomers were tried. These only gave a small reduction in shake, whilst increasing noise and vibration transmission into the cabin. It was evident that to control powertrain modes without sacrificing refinement, significantly increased damping was required, though only at the discrete powertrain rigid body mode frequencies.

The body in white package of the 400 necessitates the use of a bush on the right hand side of the vehicle. Existing hydrabushes satisfied the low frequency performance, but were found to transmit too much high frequency vibration. Avon-Clevites' proposal of incorporating a decoupler into the hydrabush was shown to meet Rovers specified stiffness characteristics.

The tune of the hydrabush, that is, the frequency and amplitude at which the primary resonance of the fluid occurs was initially determined using the PC based ride model. It was shown that minimising the response of the powertrain roll/bounce/pitch mode at 11 Hz gave the optimum benefit. For maximum energy to be dissipated at this frequency the hydrabush was specified to have a peak loss angle (45 degrees for an amplitude of $\pm$ 0.3 mm) at 9.5 Hz.

Assessment of the new 400, showed that the introduction of the hydrabush during the ride work with the new spring/damper rates, reduced the discrete band of engine shake, thereby markedly improving the secondary ride. The final hydrabush dynamic rating curves are shown in Fig 6.

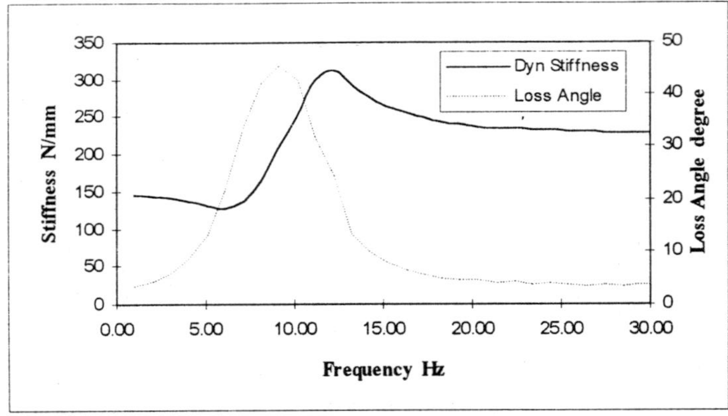

Fig 6 Hydrabush Characteristics - input amplitude ± 0.3 mm

4.4 NASA Ride Numbers

NASA ride numbers are an objective measure of occupant comfort to vibration inputs (6). The lower the number the better the comfort.

To substantiate the subjective gains of the one to one linear damping and the hydrabush, and to monitor the progress of these components through development to production, the 4 poster road simulation rig was used as an objective analysis tool in conjunction with NASA ride numbers. Fig 7 illustrates the improved ride performance of the initial and final specifications of the new Rover 400 and that of the existing 214. The table showing the corresponding NASA ride numbers, demonstrating the improvement.

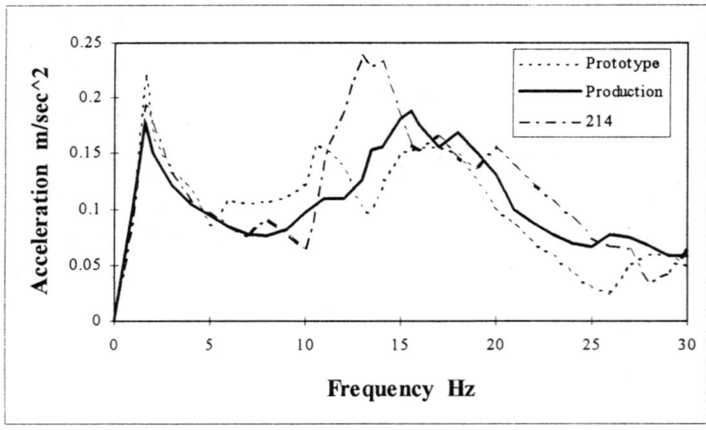

Fig 7 Comparison of seat rail response